SpringerBriefs in Molecular Science

More information about this series at http://www.springer.com/series/8898

Alexander Anim-Mensah · Rakesh Govind

Prediction of Polymeric Membrane Separation and Purification Performances

A Combined Mechanical, Chemical
and Thermodynamic Model
for Organic Systems

 Springer

Alexander Anim-Mensah
Engineering, WW-Business Unit
ITW/Hobart Corporation Technology
    Center
Troy, OH
USA

Rakesh Govind
Chemical Engineering
University of Cincinnati
Cincinnati, OH
USA

ISSN 2191-5407   ISSN 2191-5415   (electronic)
ISBN 978-3-319-12408-7   ISBN 978-3-319-12409-4   (eBook)
DOI 10.1007/978-3-319-12409-4

Library of Congress Control Number: 2014957167

Springer Cham Heidelberg New York Dordrecht London

Printed on acid-free paper

Springer is part of Springer Science+Business Media (www.springer.com)

The first author dedicates to his father Emmanuel Anim-Mensah, late mother Kate Adwoa Animah and also his little angel Xandra Anim.

Preface

There is a huge interest and need to applying polymeric membranes for separations and purifications involving organic solvents, especially solution–diffusion membranes in polar aprotic and its mixtures in industries including fine chemicals, pharmaceutical, petroleum, oil, and biotechnology. However, several drawbacks including unpredictable performances, nonuniform characterization models, lack of in-depth understanding for proper system selection, excessive swelling and compaction, irreversible fouling and membranes instability in these environments, among others, have been very deterring. Not much has been done in this area where polymeric pressure-driven solution-diffusion membranes are used for separations involving these organic solvent(s). Moreover, some of the existing models used to characterize polymeric solution-diffusion membranes such as Nanofiltration (NF), Reverse Osmosis (RO), and Pervaporation (PV) in the organic solvents reviewed are deficient in the necessary parameters for characterization; the reasons could include unavailability of necessary published data.

The complexity includes the numerous parameters to be considered in characterization models and how to combine them in a model. Polymeric membranes in these environments, in order to achieve the right performance, are exposed to considerable swelling and/or compaction simultaneously, while membrane is constrained and permeated. This brings into effect the interplay of combined chemical, mechanical, thermodynamics considerations while transporting across these membranes under a driving force. Moreover, depending on the type of organic solvent, membrane material, solute type, applied pressure, and operating conditions there is a difficulty in predicting the performance of the membranes.

My motivation came from the quest to develop a combined chemical, mechanical, thermodynamics model that represents these membranes reliably and provides in-depth explanations to assist in predicting their behaviors and performances for proper selection. In addition, I wanted to use graduate school experiences in organic solvent system separations and application of acoustics, coupled with my industrial experiences, which has prepared me sufficiently to contribute my ideas to this book to bridge some of the gaps lacking in this area.

Weber et al.'s theoretical model for describing membrane constraint in polymer-electrolyte fuel cells provided some insight into developing the model presented in this book on pressure-driven solution-diffusion polymeric membranes applied for separations and purifications involving organic solvents.

This book consisting of six chapters is designed to meet the need of membrane researchers, scientists, and engineers in academia and/or industry, who seek in-depth understanding on factors affecting solution-diffusion membranes, their behaviors, and performances. Chapter 1 presents the introduction, Chap. 2 the background to support the model development, while Chap. 3 discusses how key parameters are correlated in a model with in-depth explanations of their effects on performance and behavior. Moreover, in-depth discussions are provided on effects of combined swelling and compaction on membrane performance. Chapter 4 is on model application to published information while Chap. 5 presents a summary on the key take away or learnings from the book. Chapter 6 is on the future directions since there are a lot more areas to apply the model and extend it to several areas. A spin-off from the developed model in this book led to the definition of a new membrane dimensionless number characteristic of the separation system which will be explored in the future.

Acknowledgments

The first author's appreciation goes to the former MAST Center that was co-centered at the University of Cincinnati, Cincinnati-Ohio and University of Colorado, Boulder-Colorado for exposure to polymeric Nanofiltration (NF) and Reverse Osmosis (RO) membrane applied to aqueous, aqueous-organic, and purely organic solvent separation systems. Moreover, for the exposure to the use of Ultrasonic Time-Domain Reflectometry (UTDR), i.e., acoustics for real-time analysis and measurements of membrane swelling and compaction while permeated. UTDR is a tool for real-time visualizing of membrane behaviors and determining membrane mechanical properties while permeated, which is useful for obtaining some of the information and parameters used in this book. Also, thanks go to Prof. Shamsuddin Ilias of the Department of Chemical, Biological and Bio Engineering, North Carolina A&T State University, NC for introducing me to ceramic (inorganic) Microfiltration (MF) and Ultrafiltration (UF) membranes applied for liquid and supercritical CO_2 separations. Finally, to Prof. Stephen J. Clarson, former Director of MAST and currently at the Department of Chemical, Biomedical and Environmental Engineering, University of Cincinnati, Ohio who has always urged me to present my ideas into a book.

Contents

About the Authors

Dr. Alexander Anim-Mensah has a Ph.D., M.Sc., and B.Sc. in Chemical Engineering from the University of Cincinnati-Ohio, North Carolina A&T State University-NC, and the University of Science and Technology-Ghana, respectively. His area of specialization includes membranes science and technology, process design and engineering, water treatment, water chemistry, energy and environmental management, project economics and feasibility analysis, modeling, and cost estimation. Currently, Dr. Anim-Mensah is an Engineering Manager at ITW-Hobart Corporation, USA where he is involved in managing and leading advanced engineering, energy recovery, and R&D projects. He has authored a book on Nanofiltration membranes for organic solvent separation, co-authored two American Chemical Society book chapters on Silicone Materials, five peer-reviewed scientific articles, ten scientific conference proceedings, and presented over 15 national and international conferences. Currently, he serves in the following capacity: (1) External Advisory Board Member for Biomedical, Chemical and Environmental Engineering Program at the University of Cincinnati-OH, (2) Director of Industrial Partnership for African Membrane Society-USA, (3) Advisory Board Member for Clean WaterNet, Cleveland-OH, (4) Board Member for Alliance for Progressive Africa, Cincinnati-OH, and (5) Executive Safety Member at ITW-Hobart Corporation, Troy-OH.

Professor Rakesh Govind Ph.D., joined The University of Cincinnati in 1979 and has taught Chemical Engineering courses for over 25 years. He graduated from Indian Institute of Technology (Kanpur) with a B.Tech. Degree in Chemical Engineering and completed his MS and Ph.D. in the Process Synthesis area from Carnegie Mellon University in Pittsburgh, PA. Dr. Govind was appointed Director of the Industrial Control and Process Safety Center at Mellon Institute and worked as Senior Engineer at Polaroid Corporation, Boston, MA. He has published over 100 peer-reviewed papers, has been awarded the Alfred Bodine Award from the Society of Manufacturing Engineers, and several Earth Day Awards by the Cincinnati Gas and Electric Company. Dr. Govind served as an invited member at the Scientists Helping America Conference, convened by the Department of Defense after the 9/11 attacks. His background includes membrane processes, waste management, water treatment, simulations, and process design.

Abbreviations

NF   Nanofiltration
PV   Pervaporation
RO   Reverse Osmosis
UTDR Ultrasonic Time-Domain Reflectometry

Nomenclature

C_{bi}	Solute concentration of species i in feed solution
C_{pi}	Solute concentration of species i in permeate
Cs	Solute concentration of species i at the membrane surface
C_m	Solute concentration of species i in the membrane
PH	High pressure side
PL	Low pressure side
ΔP (i.e., PH-PL)	Transmembrane pressure
L	Membrane thickness
x	Distance along membrane thickness
K_d	Distribution or partition coefficient of the solute in the permeate (Cpi) to the solute in the membrane (cm)
D	Diffusivity
Q	Flux
δ	Concentration polarization layer
Lp	Hydraulic permeability
φ	Osmotic coefficient
R	Molar gas constant
T	Temperature
σ'	Reflection coefficient
R_i	Observed rejection of species i
r_i	Intrinsic rejection of species i
σ	Compressive stress
ε	Compressive strain
ε	Individual compressive strain
E_t	Compressive Young's modulus
E_i	Individual compressive Young's modulus
ν	Poisson ratio
V_o	Membrane dry volume
V_f	Membrane freely swollen volume
V_c	Constraint swollen membrane volume
λ_f	Freely swollen membrane solvent content

λc	Constrained swollen membrane solvent content
λp	Compacted swollen membrane solvent content
$\underline{\xi c}$	Degree of constraint of a free swelling membrane
$\overline{V}_m$	Partial molar volume of the membrane
$\overline{V}_o$	Partial molar volume of the solvent
μ_i	Chemical potential of species i
μ_f	Chemical potential relating to unconstrained or freely swelling system
μc	Chemical potential relating to constrained system
EW	Equivalent membrane weight
M_i	Molality of solute i in the membrane
K	Bulk Modulus
τ	Dilatation stress
ρ_i	Density of species i
δ_i	Solubility parameter of species i
$\Delta\delta$	Solubility parameter difference
χ	Interaction parameter
L_f	Swollen membrane thickness
Lc	Compacted membrane thickness
L_s	Total membrane thickness after swelling
β	Membrane dimensional parameter
θ	Ratio of compacted membrane thickness to swollen membrane thickness
$\alpha_{org/water}$	Separation factor of organics over water
$\eta_{water/org}$	Separation factor of water over organics
$\rho/\Delta\delta^2$	Is a parameter used as approximated Cpi for trend prediction of an aqueous- organic systems
ψ	Is a parameter used as approximated Cpi for trend prediction of a purely organic systems

Chapter 1
Introduction

Abstract This chapter presents the background behind the book with key considerations as well as what the book seeks to address as well as readers who will benefit from this book. In addition, it presents the outline of the whole book as well as what each chapter presents.

Keywords Separation · Membrane researchers · Engineers · Manufacturers · Solvent-resistant · Polymeric membranes · Solute distribution · Solution-diffusion · Reverse osmosis · Nanofiltration · Pervaporation

This book is designed to meet the need of membrane researchers, scientists, and engineers in academia and/or industry who seek in-depth understanding in separations and purifications of solution–diffusion membranes such as nanofiltration (NF), reverse osmosis (RO), and pervaporation (PV), their behaviors and performances involving organic solvents.

This book has six (6) chapters. Chapter 1 presents the introduction; Chap. 2 presents the background information to support the discussions in the book and the development of a combined chemical, mechanical, and thermodynamic model for characterization, while Chap. 3 is on each parameter correlate to performance with in-depth explanations on each and coupled parameter effects on performance. In-depth discussion is provided on effects of combined swelling and compaction on membrane performance. Chapter 4 is on developed model application to predict the trends of published information on PV and NF membrane performances involving polar protic and polar aprotic solvents, while Chap. 5 presents a summary on the key take away from book. Chapter 6 is on the future directions which includes exploring a newly defined membrane dimensionless number characteristic of the separation system.

Parameter used in this book for characterizing and predicting these membrane performances includes the following: the interactions between the membrane, solvent, and solute; solute distribution between the solvent and the membrane (K_d); solubility parameter differences between the solvent, membrane, and solute

© The Author(s) 2015
A. Anim-Mensah and R. Govind, *Prediction of Polymeric Membrane Separation and Purification Performances*, SpringerBriefs in Molecular Science, DOI 10.1007/978-3-319-12409-4_1

(i.e., $\Delta\delta = \delta_i - \delta_j$), membrane compressive Young's modulus (E) while permeated with the solvent or solution of interest, Poisson's ratio (v) relating to membrane densification, solvent density (ρ_i), molar volumes of the solvent $\left(\overline{V_0}\right)$ and membrane material $\left(\overline{V_m}\right)$, membrane constraint swelling (L_f), and compacted thickness (L_c) under applied transmembrane pressure.

1.1 General Overview on Polymeric Solution–Diffusion Membranes in Organic Solvents

Huge interests exist for separations and purifications involving organic solvents using membrane technologies in industries including the fine chemicals, pharmaceutical, petroleum, oil, and biotechnology. However, several drawbacks exist which this book presents information useful for understanding when these membranes are in these environments and how to predict their performance and/or select them. Some of the drawbacks including unpredictable performances, nonuniform characterization models, lack of in-depth understanding for proper system selection, excessive swelling and compaction, irreversible fouling, and membranes instability in these environments among others have been very deterring.

No very much information exists on polymeric membranes used in the nonaqueous environments especially polar aprotic solvents. This is because nonaqueous membrane separation systems have several sensitive parameters that affect their performances. This includes the possible strong interactions that exist between the membrane, solvent, and solute among others as well as the operating conditions. The strong interactions that could exist between the membranes and solvents could result in excessive swelling and/or compaction which lead to low separation performances if the systems are not selected properly [1, 2]. Moreover, excessive swelling leads to mechanical instability, while membrane is in operation. In some cases, this interaction eventually results in membrane degradation. Another possibility is accelerated fouling and irreversible fouling as results of affinity between organic solutes and polymeric membranes if membrane materials are not selected properly. Additionally, strong interactions between the solvent and solute could lower the membrane ability to reject the solute. From the foregoing information, it is clear that the interactions that existing between the solvent, solute, and membrane for organic systems require careful understanding to select these membranes to prevent lowered separation efficiencies. Swelling and compaction is expected for solution–diffusion membranes such as RO, NF, and PV membranes operated in organic systems [1, 2]. However, prevention of excessive swelling, compaction, and irreversible fouling makes the consideration of chemical, mechanical, and thermodynamic properties important in characterizing and selecting membranes involving organic separations.

The complexities encountered when characterizing these membranes include the consideration of numerous parameters, how to combine these parameters into a model, and availability of published data. Polymeric membranes in these environments to achieve the right performance are exposed to simultaneously considerable swelling and/or compaction, while membrane is constrained and permeated. This brings into effects the interplay of combined chemical and mechanical considerations while transporting cross these membranes under a driving force. Moreover, depending on the type of organic solvent, membrane material, solute type, applied pressure, and operating conditions, there is a difficulty in characterizing and predicting these membrane performances reliably. This book seeks to address some of the major issues present a model developed with the necessary parameters for providing insights to assist understanding and selecting these membranes reliably.

References

1. Anim-Mensah, A. R. (2007). Evaluation of solvent resistant nanofiltration (SRNF) membranes for small-molecule purification and recovery of polar aprotic solvents for re-use (Ph.D. thesis, University of Cincinnati, OH, 2007).
2. Anim-Mensah, A. R. (2012). *Nanofiltration membranes assessment for organic systems separations*. Germany: LAP LAMBERT Academic Publishing.

Chapter 2
Background

Abstract The information necessary to support the model development and the literature to back the various discussions in the book is presented here. This chapter presents information on the different organic solvents and their properties, membrane materials and mechanical properties as well as the necessary membrane terminologies, and information on the model that provided insight into developing the model developed in this book.

Keywords Organic solvents · Membranes · Membrane performance · Membrane polymers · Membrane models · Membrane swelling · Membrane compaction · Young's modulus · Solubility parameter · Poisson ratio · Membrane rejection · Solvent-resistant · Membrane characterization · Solute distribution · Stress · Strain

2.1 Organic Solvents

Applications involving organic solvents is on the rise with a 2013 annual sales of 25 billion USD. Moreover, experts predict a 4 % rise in annual sales until 2021 [1]. Generally, organic solvents are grouped into polar and nonpolar solvents. Polar solvents are subcategorized into polar parotic and polar aprotic (Fig. 2.1 and Table 2.1). Table 2.1 presents examples of nonpolar, polar parotic, and polar aprotic solvents. Nonpolar solvents include toluene, hexane, and benzene [1]. Polar parotic have hydrogen atom attached to an electronegative atom such as oxygen and include water, ethanol, and acetic acid. Polar aprotic solvents have large dipole moments and dielectric constants compared with nonpolar solvents. They are very aggressive with their molecules having bonds that include multiple and large bond dipole. These multiple bond exists between carbon (C), nitrogen (N), sulfur (S), phosphorus (P), or oxygen, such as in ketones, aldehydes, ethyl acetate, and dimethylformamide (DMF) [2].

5

A. Anim-Mensah and R. Govind, *Prediction of Polymeric Membrane Separation and Purification Performances*, SpringerBriefs in Molecular Science,
DOI 10.1007/978-3-319-12409-4_2

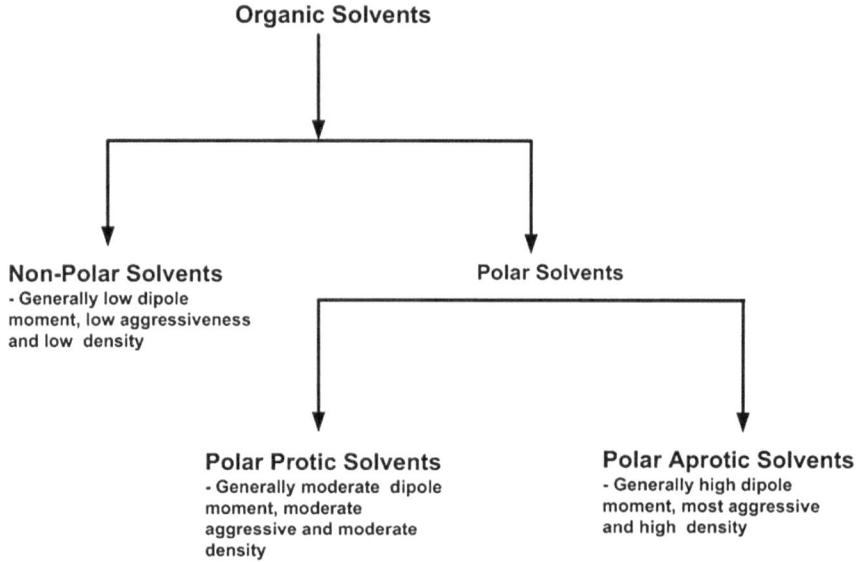

Fig. 2.1 General grouping of organic solvents [2]

Table 2.1 Properties of some organic solvents [3–5]

Solvents	Dielectric constant	Density (g/ml)	Dipole moment (D)	Solubility parameter $(MPa^{1/2})$
Nonpolar solvents				
Pentane	1.84	0.626	0.00	14.5
Cyclopentane	1.97	0.751	0.00	16.6
Hexane	1.88	0.655	0.00	14.9
Cyclohexane	2.02	0.779	0.00	16.8
Benzene	2.30	0.879	0.00	18.8
Toluene	2.38	0.867	0.36	18.2
1,4 dioxane	2.30	1.033	0.45	20.7
Chloroform	4.81	1.498	1.04	18.8
Diethyl ether	4.30	0.713	1.15	15.1
Polar protic solvents				
Formic acid	58	1.210	1.41	24.9
n-Butanol	18	0.810	1.63	23.1
Isopropanol	18	0.785	1.66	24.9
n-Propanol	20	0.803	1.68	24.5
Ethanol	24.55	0.789	1.69	26.0
Methanol	33.00	0.791	1.70	29.7
Acetic acid	6.20	1.049	1.74	21.4
Water	80.00	1.000	1.85	47.9

(continued)

Table 2.1   (continued)

Solvents	Dielectric constant	Density (g/ml)	Dipole moment (D)	Solubility parameter (MPa$^{1/2}$)
Polar aprotic solvents				
Dichloromethane	9.1	1.327	1.60	20.0
Tetrahydrofuran	7.5	0.886	1.75	18.6
Ethyl acetate	6.02	0.894	1.78	18.2
Acetone	21	0.786	2.88	19.7
Dimethylformamide	38	0.944	3.82	24.8
Acetonitrile	37.5	0.786	3.92	24.7
Dimethyl sulfoxide	46.7	1.092	3.96	24.5
N-Methyl pyrrolidinone	32.2	1.028	4.1	22.9
Propylene carbonate	64	1.205	4.90	27.2

2.2   Some Membrane Performance Parameters and Models

For a membrane undergoing permeation, the various solute concentrations involved are the bulk solute concentration C_b, the permeate solute concentration C_{pi}, the membrane surface solute concentration C_s, and the solute concentration in the membrane C_m that are shown in the Fig. 2.2.

Note that C_m varies across the membrane thickness. For simplicity C_m at any point x along the membrane thickness can be calculated using $C_{mx} = C_s - (C_s - C_{pi})x/L$. For average solute concentration in the membrane, $x/L = 0.5$ could be used. C_m variation with the membrane thickness for each separation system is different so as the

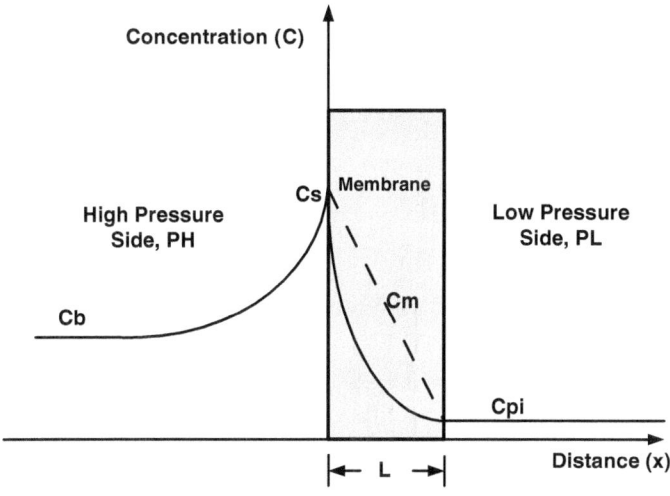

Fig. 2.2   Membrane under permeation solute concentrations

ratio of the C_{pi} to the C_m (i.e., $C_{pi}/C_m = K_d \leq 1$) at the various transmembrane pressures and feed rates. C_{pi}, C_s, and C_b are related to the flux q across the membrane, diffusivity D, and the concentration polarization layer δ as shown in Eq. 2.1 [2, 6, 7].

$$\frac{\delta}{D}q = \ln\left(\frac{C_s - C_{pi}}{C_b - C_{pi}}\right) \tag{2.1}$$

δ/D in Eq. 2.1 is obtained from Eq. 2.2 by knowledge of flux, transmembrane pressure ΔP, osmotic coefficient φ_m, solute concentrations in the bulk feed C_b and permeate C_{pi}, molar gas constant R, and temperature T. The hydraulic permeability L_p and the reflection coefficient σ^1 are also obtained by fitting Eq. 2.2 to experimental data [7, 8].

$$\ln\left[\frac{\Delta P - \frac{q}{L_p}}{\varphi_m RT\left(C_b - C_p\right)}\right] = \ln y = \frac{\delta}{D}q + \ln\sigma^1 \tag{2.2}$$

The basic parameters used to describe membrane performances include flux (q) and observed rejection (R_i) of species i [9]. The observed rejection is defined in Eq. 2.3;

$$\text{Observed Rejection } (R_i) = \left(1 - \frac{C_{pi}}{C_b}\right) \times 100\,\% \tag{2.3}$$

Equation 2.3 shows that an increase in C_{pi} results in a decrease in the objection rejection, i.e., membrane performance. Another parameter of interest is intrinsic rejection (r_i) of species i defined to take into consideration the concentration polarization layer C_s on the high pressure side of the membrane is as in Eq. 2.4 [9].

$$\text{Intrinsic Rejection } (r_i) = \left(1 - \frac{C_{pi}}{C_s}\right) \times 100\,\% \tag{2.4}$$

where C_{pi}, C_b, and C_s are the permeate solute concentration, bulk feed solute concentration, and the solute concentration at the surface of the membrane at the high pressure side, respectively. For a membrane retaining solute effectively, C_s is far greater than C_b. Membrane flux (q) is defined as in Eq. 2.5 [9];

$$\text{Flux } (q) = \frac{\text{Volume of Permeate}}{\text{Membrane Area} \times \text{Time}} \tag{2.5}$$

2.3 Some Factors Affecting Membrane Performance

Factors affecting polymeric membrane performance in organic systems include concentration polarization [9], swelling [10], compaction [11], fouling [12], and affinity between the solvent, solute, and membrane. Swelling and compaction are known to have negative effects on performance; however, they have some positive effects. Concentration polarization and membrane surface fouling could create a thin resistance layer that could improve membrane rejection, however, at the expense of flux reduction in some cases. Swelling of membrane polymer by a solvent may indicate closeness in the individual solubility parameters. Swelling results in flux increase; however, excessive swelling could lead to excessive compaction which may lead to lower flux [2]. In addition, excessive swelling could reduce membrane rejection. Moreover, compaction reduces the membrane thickness which reduces the diffusion path resulting in increased flux; however, the increase compaction could also lead to increased membrane resistance leading to flux drop. Mechanically, excessive membrane swelling could result in low Young's' modulus since the strain could be high at a low applied transmembrane pressure. Reasonable transmembrane pressure on a swollen membrane could lead to low membrane rejection but high flux [11] since the swollen membrane network could transport the solutes easily across the membrane. However, in some situation, very high transmembrane pressure on slightly swollen membrane could lead to increase rejection but low flux if the concentration polarization builds up quickly.

2.4   In-Situ Real-time Swelling and Compaction Measurement

Ultrasonic time domain reflectometry (UTDR) are among the technologies which could be used to measure membrane swelling as well as compaction while the membrane is permeated with the solvent or solutions of interest in real time [2, 13]. Details of UTDR are found elsewhere [2].

2.5   Description of Material Mechanical Properties

Some parameters used to describe polymer mechanical properties depending on the application include the Young's (tensile or compressive), bulk, shear, and flexural moduli, tensile or compressive strength, yield stress and tensile or compressive strain and failure, and Poisson ratio. Most mathematical models use some of the above parameters in constitutive equations for descriptions and predictions of mechanical systems [14]. Table 2.2 shows some solvent resistant membrane materials with their respective Poisson ratios, Young's moduli (tensile and compressive), and solubility parameters and densities. Note that the parameters in Table 2.2 were obtained when the materials surfaces were unconstrained and under tension or compression. For pressure-driven solution–diffusion membranes, the surfaces are generally constrained and compacted while permeated at a given transmembrane pressure (compressive stress) and feed rate.

Figure 2.3 shows a homogenous membrane with initial thickness L_D, compacted while permeated at a transmembrane pressure (High HP–Low pressures LP). Homogenous membranes are made of one material and have either a dense or porous structure.

Table 2.2 Mechanical and physicochemical properties of some candidate solvent resistant membrane polymers obtained under unconstrained conditions [16–25]

Solvents	Tensile Young's modulus (E) (MPa)	Compressive Young's modulus (E) (MPa)	Poisson ratio (ν)	Solubility parameters δ (MPa$^{1/2}$)	Density (kg/dm^3)
Polydimethylsiloxane (PDMS)	0.27–0.83	0.56–3.59	0.5	14.9–15.59	0.95–1.25
Poly(1-(trimethylsilyl)-1-propyne) (PTMSP)	630	–	–	–	0.964
Cellulose acetate	2,100–4,100	–	0.40	21.9–27.8	1.29–1.30
Polyimide	2,000–3,000	2,500–4,100	0.34–0.42	–	1.42
Polyester	200–400	2,737–2,848	0.12–0.35	–	1.00–1.42
Polyphenylsulfone (PPSU)	2,340–2,480	1,730–1,931	0.42	20	1.29
Poly(ether ether ketone) (PEEK)	2,700–12,000	3,450–4,140	–	21.2–22.6	1.23–1.50
Polysulfone (PS)	2,480–8,690	2,580–8,000	0.37–0.42	20	1.24
Polyamide	1,500–16,000	2,241–2,896	0.30–0.50	22.87–27.8	1.06–1.39
Polyacylonitrile (PAN)	5,200		–	31.5	1.18–1.38
Polyether ketone (PEK)	3,190		–	20	1.27–1.43
Polyarylene ether ketone	–	–	–	–	–
Polyether sulfone (PES)	2,600–6,760	2,680–7,720	0.41	22.9–23.12	1.37–1.60
Polyetherimide (PEI)	2,900–3,447	2,900–3,300	0.36	–	1.28
Polyphenylene sulfide (PPS)	3,200–14,000	2,965	–	–	1.30–1.64
Poly(amide–imide)	4,500–18,600	4000–9900	0.39–45	–	1.38–1.42

Fig. 2.3 A homogenous
membrane under compression

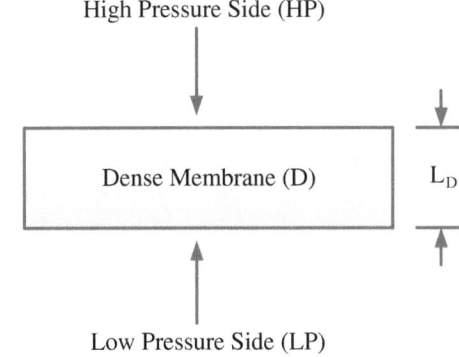

Fig. 2.4 An asymmetric
with skinned functional
layer membrane under
compression

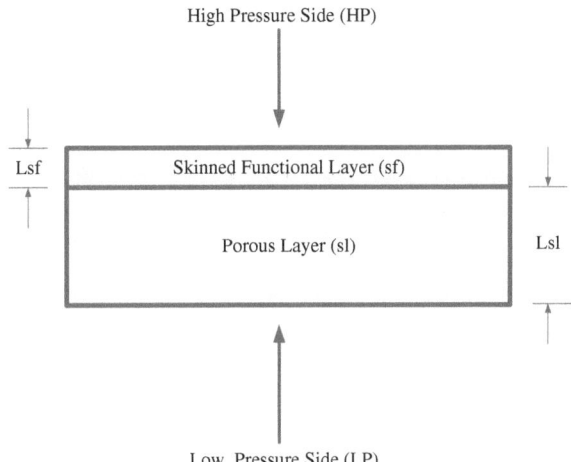

Figures 2.4 and 2.5 show asymmetric membranes with two (2) layers, i.e., a functional and support layers. Figure 2.4 shows an asymmetric membrane made of the same material but has a skinned functional layer (sf) and a support layer (sl); hence, different porosities or pore structures permeated at a transmembrane pressure.

Figure 2.5 shows an asymmetric membrane which the various layers, i.e., functional (f) and support layers (s) could be of different materials, hence different porosities or pore structures as in a composite membrane and permeated at a transmembrane pressure.

For a homogenous membrane (Fig. 2.3) under compression while permeated at a given transmembrane pressure same as compressive stress (σ), the compressive Young's modulus E_t is affected by the membrane material, porosity, and solvent–membrane interaction. The compressive Young's modulus E_t is obtained from the linear portion of the plot of the compressive stress (σ) or the transmembrane pressure versus the strain (ε) as shown in Eq. 2.6 [15];

Fig. 2.5 An asymmetric with
a functional layer membrane
under compression

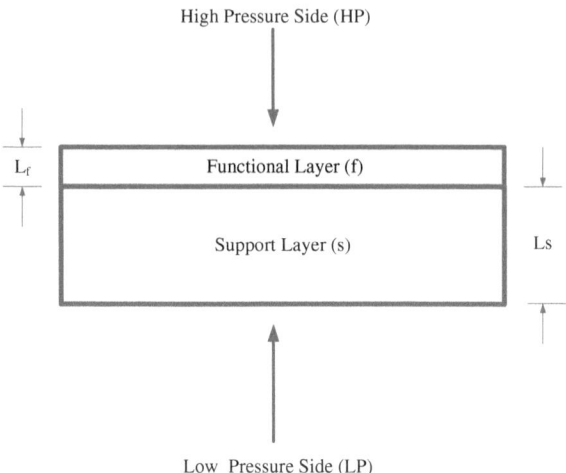

$$E_t = \frac{\text{Compressive Stress } (\sigma)}{\text{Strain } (\varepsilon)} \tag{2.6}$$

For an asymmetric membrane under compression while permeated at a given transmembrane pressure, each of the layers undergoes different strains. Let the stain for an asymmetric membrane with the functional layer (f) or the skin functional layer (sf) have with corresponding strains of ε_f while that of the support layer (s) or porous layer (sl) have a strain of ε_s. The individual compressive Young's modulus of each of the layers can be calculated as shown in Eq. 2.7 [15];

$$E_f = \frac{\sigma}{\varepsilon_f} \quad \text{and} \quad E_s = \frac{\sigma}{\varepsilon_s} \tag{2.7}$$

From knowledge of the individual compressive Young's modulus, the composite compressive Young's modulus can be calculated using Eq. 2.8 [15];

$$E_t = \frac{E_f E_s}{E_f L_f + E_s (1 - L_f)} \tag{2.8}$$

where $L_s (L_{sf}) + L_f (L_{sl}) = 1$ [15].

However, for a membrane made of layers undergoing permeation, it will be difficult to obtain the strain of the individual layers; hence, an overall strain will be necessary to obtain the composite compressive Young's modulus.

Most materials resist change in volume more than change in shape and are being defined by bulk and shear moduli, respectively. For most isotopic and elastic materials with unconstrained surfaces, the Poisson ratio (ν) is in the range of -1 to 0.5 (i.e., $-1 \leq \nu \leq 0.5$) for stability [26, 27]. However, situations including densification, anisotropism, and constrained surfaces could lead to material Poisson ratios out of the normal range ($-1 \leq \nu \leq 0.5$) while material is still stable. Materials with different Poisson ratio behave mechanically different [27]. For pressure-driven solution–diffusion membranes, the surfaces are constrained and compacted while permeated at a given transmembrane pressure (i.e., compressive stress) and feed rate. For perfectly constrained membranes, the transverse strain as results of swelling and compaction is far higher that the longitudinal strain since the membrane is constrained. This leads to membrane densification which results in higher than expected Poisson ratios values. Hence, the Poisson ratios could be very well outside of the normal ranges for membranes in operation.

2.6 Constraint Polymeric Membrane Model

Weber and Newman [28] defined Eqs. 2.9–2.18 in their theoretical study of constraint polymer-electrolyte fuel cells and compared free swelling and constraint swelling polymer membrane (see Fig. 2.6).

For a free swelling membrane;

$$V_f = V_o \left(1 + \frac{\lambda_f \overline{V_o}}{\overline{V_m}} \right) \tag{2.9}$$

For a constraint swelling membrane;

$$V_c = V_o \left(1 + \frac{\lambda_f \overline{V_o}}{\overline{V_m}} (1 - \xi_c) \right) \tag{2.10}$$

where ε_c is the degree of constraint defined by

$$\xi_c = \frac{V_f - V_c}{V_f - V_o} \tag{2.11}$$

where V_o, V_f, and V_c are the initial dry membrane volume, free swollen membrane volume, and the constraint swollen membrane volume, respectively. λ_f, $\overline{V_m}$, $\overline{V_o}$ and ξ_c are the unknown average membrane solvent content, partial molar volume of the membrane and solvent, and the degree of constraint of a free swelling membrane, respectively.

For a membrane undergoing free and constraint swelling, Weber and Newman [28] defined the chemical potential internal and external of the membrane to be equal at equilibrium in Eqs. 2.12–2.14.

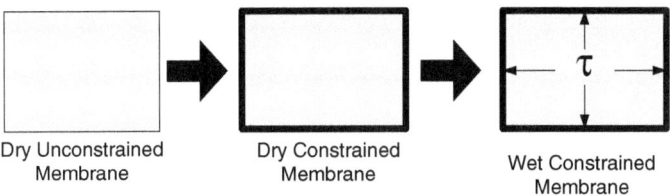

Fig. 2.6 Membrane undergoing constrain swelling [28]

$$\mu_{f,c}^{int} = \mu_{f,c}^{ext} \tag{2.12}$$

$$\mu_f = \mu^* + RT \ln \lambda_f EW + 2RT \sum_{j=1}^{n} \chi_{i,j}^* M_j \tag{2.13}$$

$$\mu_c = \mu^* + RT \ln \lambda_c EW + 2RT \sum_{j=1}^{n} \chi_{i,j}^* M_j + \overline{V_o} \tau \tag{2.14}$$

$$\mu_f = \mu_c \tag{2.15}$$

$$\tau = \frac{RT}{\overline{V_o}} \ln\left(\frac{\lambda_c}{\lambda_f}\right) \tag{2.16}$$

$$\tau = -K \ln\left(\frac{V_c}{V_f}\right) \tag{2.17}$$

$$\left(\frac{\lambda_c}{\lambda_f}\right) = \left(\frac{\overline{V_m} + \overline{V_o}(1 - \xi_c)\lambda_f}{\overline{V_m} + \overline{V_o}\lambda_f}\right)^{\frac{K\overline{V_o}}{RT}} \tag{2.18}$$

where μ_{fc}^{int} and μ_{fc}^{ext} are the chemical potential for free (f) and constraint (c) relative to the internal and external of the membrane. μ^* is the reference chemical potential. EW, M_j, K, and τ are the equivalent membrane weight; molality of the solute j in the membrane, K is the bulk modulus and dilatation stress, respectively. R and T are the molar gas constant and the temperature, respectively.

References

1. Ceresana Market Intelligence Consulting, (2014). Market Stud: Solvents (3rd ed.). http://www.ceresana.com/en/market-studies/chemicals/solvents/
2. Anim-Mensah, A. R. (2007). Evaluation of solvent resistant nanofiltration (SRNF) membranes for small-molecule purification and recovery of polar aprotic solvents for re-use (Ph.D. thesis, University of Cincinnati, OH, 2007).
3. Lowery, T. H., & Richardson, K. S. (1987). *Mechanism and theory in organic chemistry* (3rd ed.). San Francisco: Benjamin-Cummings Publishing Company.
4. Barto, A. F. M. (1991). *CRC handbook of solubility parameters and other cohesion parameters* (2nd ed.). Boca Raton: CRC Press LLC.
5. Hansen, C. M. (2007). *Hansen solubility parameters: A user's handbook*. Boca Raton: CRC Press LLC.
6. Anim-Mensah, A. R. (2012). *Nanofiltration membranes assessment for organic systems separations*. Germany: LAP LAMBERT Academic Publishing.
7. Anim-Mensah, A. R., Krantz, W. B., & Govind, R. (2008). Studies on polymeric nanofiltration-based water softening and the effects of anion properties on the softening process. *European Polymer Journal, 44*, 2244–2252.

8. Gupta, V. K., Hwang, S. T., Krantz, W. B., & Greenberg, A. R. (2007). Characterization of nanofiltration and reverse osmosis membrane performance for aqueous salt solutions using irreversible thermodynamics. *Desalination, 208*, 1–18.

9. Koros, W. J., Ma, Y. H., & Shimidzu, T. (1996). Terminology for membranes and membranes processes. *Journal of Membrane Science, 120*, 149–159.

10. Hicke, H. G., Lehmann, I., Malsch, G., Ulbricht, M., & Becker, M. (2002). Preparation and characterization of a novel solvent-resistant and autoclavable polymer membrane. *Journal of Membrane Science, 198*, 187–196.

11. Persson, K. M., Gekas, V., & Trägårdh, G. (1995). Study of membrane compaction and its influence on ultrafiltration water permeability. *Journal of Membrane Science, 100*, 155–162.

12. Tarnawski, V. R., & Jelen, P. (1986). Estimation of compaction and fouling effects during membrane processing of cottage cheese whey. *Journal of Food Engineering, 5*, 75–90.

13. Anim-Mensah, A. R., Franklin, J. E., Palsule, A. S., Salazar, L. A., & Widenhouse, C. W. (2010). Characterization of a biomedical grade silica-filled silicone elastomer using ultrasound, advances in silicones and silicone-modified materials. *ACS Symposium Series, 1051*, 85–98. (Chap. 8).

14. Gardner, S. H. (1998). An investigation of the structure-property relationships for high performance thermoplastic matrix, carbon fiber composites with a tailored polyimide interphase (Ph.D. dissertation, Virginia Polytechnic Institute and State University, 1998).

15. Harris, B. (1999). *Engineering composite materials*. London: The Institute of Materials.

16. Patel, M. C., & Shah, A. D. (2002). Poly(amides–imides) based on amino end-capped polyoligomides. *Oriental Journal of Chemistry, 1*, 19.

17. Charlier, C. (2012). *Study of materials, polymer data*. HELMo-Gramme Institute Belgium. Retrieved August 6, 2012, from http://www.gramme.be/unite4/Study%20of%20Materials/Polymers%20Annex.pdf.

18. Mark, J. E. (1999). *Polymer data handbook* (3rd ed.). Oxford: Oxford University Press Inc.

19. Dupont Kapton Polyimide Film General Specifications, Bulletin GS-96-7. Retrieved August 6, 2012, from http://www.dupont.com/kapton/general/H-38479-4.pdf.

20. Stafie, N. (2004). Poly(dimethylsiloxane)-based composite nanofiltration membranes for non-aqueous applications (Ph.D. dissertation, University of Twente, The Netherlands, 2004).

21. Technical Data, Solvay Specialty Polymer. Retrieved August 6, 2012, from http://www.solvayplastics.com/sites/solvayplastics/EN/specialty_polymers/Pages/solvay-specialty-polymers.aspx.

22. Burke, J. (1984). *Solubility parameter: Theory and application* (Vol. 3, 13–58). AIC Book and Paper Group Annual.

23. Matweb, Material Property Data. Retrieved August 6, 2012, from www.matweb.com.

24. Strong, A. B. (2008). *Fundamental of composite manufacturing materials, methods and applications* (2nd ed.). Dearborn: Society of Manufacturing Engineering.

25. Wang, Z. (2011). Polydimethylsiloxane mechanical properties measured by macroscopic compression and nanoindentation techniques (Ph.D. thesis, University of South Florida, 2011).

26. Fung, Y. C. (1968). *Foundation of solid mechanics*. Englewood, NJ: Prentice-Hall.

27. Greaves, G. N., Greer, A. L., Lakes, R. S., & Rouxel, T. (2011). Poisson's ratio and modern materials. *Nature Materials, 10*, 823–837.

28. Weber, A. Z., & Newman, J. (2004). A theoretical study of membrane constraint in polymer-electrolyte fuel cells materials, Interfaces, and electrochemical phenomena. *AIChE Journal, 50*, 3215–3226.

Chapter 3
Model Development and Effects of the Various Model Parameters

Abstract The step-by-step process involved in developing the membrane performance prediction model is presented here. In the developed model, the membrane performance measured as the concentration of the solute in the permeate is related to the Young's modulus of the membrane material while under permeation, solvent density, Poisson ratio (v), extent of membrane constraint, swelling to compaction ratio, solvent and solute solubility parameter difference; solute partitioning between the solvent and membrane; solvent and membrane polymer density ratio, as well as the solvent and membrane molar volumes. This chapter presents the sensitivity of all these parameters on membrane performance. Emphasis is placed on simultaneous membrane swelling and compaction which appears to be a major issue and presents some clues to understand the separation performances of polymeric membranes applied for organic system separation. The model shows membrane constraint and densification as results of compaction controlled transport of solute molecules across polymeric membranes. Moreover, the model is capable of determining the effects of the type solvents, i.e., polar aprotic, polar protic and non-polar on separation performance. Finally, the model developed led to defining of a new dimensionless number characteristics of the separation systems. UTDR technology is employed to extract real-time useful information about the model.

Keywords Membrane models · Chemical potential · Free swelling · Constraint swelling · Compaction · Young's modulus · Solubility parameter · Membrane densification · Poisson ratio · Dimensionless number · Organic solvents · Solute distribution · Solvent-resistant · Membrane characterization · Performance prediction · Solution-diffusion · Polymer · Swelling-compaction ratio · Pervaporation · Nanofiltration · Separation · Purification

In this chapter, the step-by-step process of developing the membrane performance prediction model, the necessary selected parameter to describe an organic separation system and considerations, and the interrelation between the various selected parameters are presented. In addition, the various sections in this chapter have

© The Author(s) 2015   17
A. Anim-Mensah and R. Govind, *Prediction of Polymeric Membrane Separation
and Purification Performances*, SpringerBriefs in Molecular Science,
DOI 10.1007/978-3-319-12409-4_3

graphical illustrations to present the sensitivity of the various parameters in the model on membrane performance. Emphasis is placed on simultaneous membrane swelling and compaction which appears to be a major issue and presents some clues to understand the separation performances of polymeric membranes applied for organic separation systems.

The model developed here had insight from Weber and Newman [1] theoretical work on constrained polymeric membrane in fuel cell application (see Sect. 2.6). However, most membranes used for separation in operation are constrained, swollen, and compacted while permeated under a transmembrane pressure. The model development here led to the definition of a new membrane dimensionless number which will be explored in the future.

It is worth to point out that the compressive Young's moduli (E) and the Poisson ratio (v) required for applying the developed model here in this chapter are different from what is published in most literature. This because both E and v required here take into consideration membrane constraining, membrane porosity, solvent interaction with the membrane and membrane densification under the applied transmembrane pressure, while most of the published data exclude the conditions outlined above. Ultrasonic Time-domain Reflectometry (UTDR) technology is among the technologies with the flexibility to achieve the required parameters in real time with less difficulties.

3.1 Model Development

Most membranes used for separation in operation are constrained, swollen, and compacted while permeated under a transmembrane pressure which is compressive. Most membranes applied for separation purposes undergo simultaneous compaction with permeation while constrained. Figure 3.1 shows a constrained dry membrane exposed to an organic solvent to swell under no pressure and permeated under a transmembrane pressure ΔP, i.e., PH–PL. Note that the extent of swelling is lower for constrained membrane than the unconstrained.

For the system in Fig. 3.1, Eq. 3.1 shows how the various parameters including the chemical potential are connected to the stresses. The variables have their usual definition as from Eqs. 2.9–2.18 (see Chap. 2, Sect. 2.6).

$$\mu_p = \mu^* + RT \ln \lambda_p \, EW + 2RT \sum_{j=1}^{n} \chi_{i,j}^* M_j + \overline{V_o}(\tau - \Delta P) \tag{3.1}$$

ξp is selected such that $\tau_p = \tau_c$

$$\mu_p = \mu_c \tag{3.2}$$

From Eqs. 2.14 and 3.1, Eq. 3.3 is derived

$$\Delta P = \frac{RT}{\overline{V_o}} \ln \left(\frac{\lambda_p}{\lambda_c} \right) \tag{3.3}$$

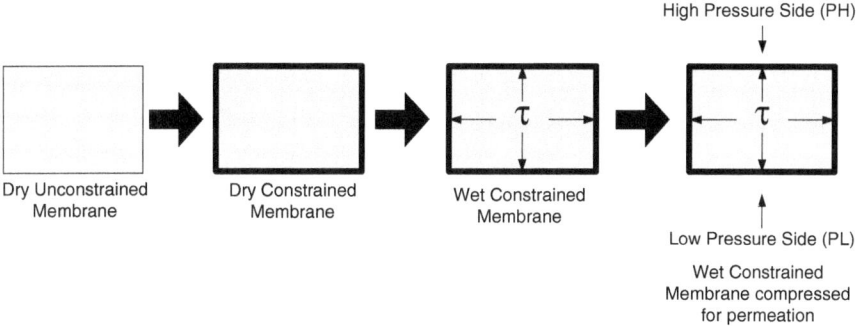

Fig. 3.1 Membrane undergoing constrain swelling and compaction under applied transmembrane pressure

From Eqs. 2.16–2.18 (see Chap. 2, Sect. 2.6); Eq. 3.4 is derived.

$$\Delta P = -K \ln\left(\frac{V_c}{V_f}\right) + \frac{RT}{\overline{V_0}} \ln\left(\frac{\lambda_p}{\lambda_c}\right) \tag{3.4}$$

But

$$\ln\left(\frac{\lambda_p}{\lambda_f}\right) = \frac{K\overline{V_0}}{RT} \ln\left(\frac{\overline{V_m} + \overline{V_0}(1 - \xi p)\lambda_f}{\overline{V_m} + \overline{V_0}\lambda_f}\right) \tag{3.5}$$

$$\xi p = \frac{V_f - V_p}{V_f - V_0} \tag{3.6}$$

$$\Delta P = -K \ln\left(\frac{V_c}{V_f}\right) + \frac{RT}{\overline{V_0}} \ln\left(\frac{\lambda_p}{\lambda_c}\right) \tag{3.7}$$

$$\Delta P = K \ln\left[\left(\frac{V_f}{V_c}\right)\left(\frac{\overline{V_m} + \overline{V_0}(1 - \xi p)\lambda_f}{\overline{V_m} + \overline{V_0}\lambda_f}\right)\right] \tag{3.8}$$

At equilibrium, consider a constraint membrane swelling in a solvent and a constraint membrane swelling in a solution comprised of one solute under pressure. Applying the condition to Eq. 2.14 (see Chap. 2, Sect. 2.6) and Eq. 3.1 and simplifying, Eq. 3.9 is defined.

$$\Delta P = \frac{RT}{\overline{V_0}} \ln\left(\frac{\lambda_p}{\lambda_c}\right) + \frac{RT}{\overline{V_0}} \chi_{i,j}^* M_j$$

$$\Delta P = \frac{RT}{\overline{V_0}} \ln\left(\frac{\lambda_p}{\lambda_f}\frac{\lambda_f}{\lambda_c}\right) + \frac{RT}{\overline{V_0}} \chi_{ij}^* M_j \tag{3.9}$$

From Eq. 2.18 (see Chap. 2, Sect. 2.6), Eqs. 3.5, 3.8, and 3.9 and expressing M_2 in terms of the other variables as in Eq. 3.10;

$$M_2 = \frac{K\overline{V_0}}{2RT\,\chi_{i,j}^*}\ln\left[\left(\frac{V_f}{V_c}\right)\left(\frac{\overline{V_m}+\overline{V_0}(1-\xi c)\lambda_f}{\overline{V_m}+\overline{V_0}\lambda_f}\right)\right] \quad (3.10)$$

Equation 3.10 becomes Eq. 3.11 after substituting the following equations and expressing C_{pi} (membrane performance) in terms of the other variables.

But; $K = \frac{E}{3(1-2v)}$ [2]; $\chi_{i,j}^* = \overline{V_0}\frac{(\delta_i-\delta_j)^2}{RT}$ [3]; $M_2 = \frac{C_m}{\rho_i}$; $K_d = \frac{C_{pi}}{C_m}$ and $\alpha = \frac{\overline{V_0}\lambda_f}{\overline{V_m}}$ but $\overline{V_m} \ggg \overline{V_0}$, i.e., molar volume of the membrane polymer $\overline{V_m}$ compared with the solvent $\overline{V_0}$. $\overline{V_i} = \frac{M_w}{\rho_i}$ M_w is the molecular weight and ρ_i is the density of species i in the membrane. λ_f is the solvent retained in a membrane and defined mathematically as $\lambda_f = \%\,\text{Swelling}\left(\frac{\rho_{\text{solvent}}}{\rho_{\text{membrane polymer}}}\right)$.

Hence $\alpha = \frac{\overline{V_0}\lambda_f}{\overline{V_m}}$ becomes $\alpha = \left(\frac{\overline{V_0}}{\overline{V_m}}\right)\left(\frac{\rho_{\text{solvent}}}{\rho_{\text{membrane polymer}}}\right)$

$$C_{pi} = \frac{10^{-6}E\rho_i K_d}{6(1-2v)(\delta_i-\delta_j)^2}\ln\left[\left(\frac{L_f}{L_c}\right)\left(\frac{1+\alpha(1-\xi c)}{1+\alpha}\right)\right] \quad (3.11)$$

10^{-6} is from units' conversion. Where K, E, and v are the bulk and Young's moduli, and the Poisson ratio, respectively; M_2 and C_m are the molality of the solute in the membrane and the average solute concentration in the membrane, respectively. χ, $\overline{V_0}$, δ_i, and δ_j are the interaction parameters of the solvent, partial molar volume of the solvent. and the solubility parameters of the solvent and solute, respectively. L_f and L_c are the membrane constrained swollen and compacted thicknesses, respectively. C_{pi} is the solute concentration in the permeate.

Most pressure-driven membranes in operation are generally near fully constrained; hence, ε_c could be approximated to 1 in Eq. 3.11 to become Eq. 3.12.

$$C_{pi} = \frac{10^{-6}E\rho_i K_d}{6(1-2v)(\delta_i-\delta_j)^2}\ln\left[\left(\frac{L_f}{L_c}\right)\left(\frac{1}{1+\alpha}\right)\right] \quad (3.12)$$

$$\Delta\delta = \left|(\delta_i-\delta_j)\right| \neq 0, \quad v \neq 1/2$$

C_{pi} cannot be negative; hence, the product of the $(1-2v)$ and $\ln\left[\left(\frac{L_f}{L_c}\right)\left(\frac{1}{1+\alpha}\right)\right]$ should results in a positive C_{pi} value. Equation 3.12 shows how some of the important variables relate to membrane performance. Equation 3.12 shows that for a separation system comprised of a membrane, solvent, and a solute, the membrane separation performance depends on the membrane compressive Young's modulus (E), Poisson ratio v, partial molar volume of the membrane $\overline{V_m}$, and dry weight of the membrane. The effect of the solute on membrane performance is through the solubility parameter δ_j. The solvent effect on the separation performance is through the solvent density ρ_i, solvent solubility parameter δ_o, and solvent partial molar volume $\overline{V_o}$. Both $\overline{V_m}$ and $\overline{V_o}$ are measure of the molecular size of the membrane polymer and the solvent. Coupled parameters include membrane–solvent interaction, and their effect on performance is shown through the membrane swelling L_f, compaction L_c, membrane–solvent content λ_f, Young's modulus E, and the Poisson ratio v. Also, the solvent–solute–membrane interaction effect is shown through $K_d(C_{pi}/C_m)$. The compressive Young's modulus for the same material will vary when in contact or permeated with different solvent because of the difference in the interactions and are different from most of the published data which excludes solvents.

Depending on the separation system, the membrane performance could either be predicted by a single dominant parameter and/or coupled parameters. The function of K_d is coupled since it is affected by the affinity between the solvent, solute, and membrane, determined by δ_i, δ_j, δ_m (membrane material solubility parameter), bulk concentration of the solute C_b, and transmembrane pressure. Hence, C_{pi} relates K_d and $(\delta_i - \delta_j)$ at a given transmembrane pressure as $C_{pi} \propto K_d/(\delta_i - \delta_j)^2$ from Eq. 3.12. Equation 3.12 shows C_{pi} relationship to the several parameters which some could be absent in many exiting published data and hence needs to be obtained through preforming the experiments yourself; hence, C_{pi} trend could be predicted on the assumption that some of the parameters are constant.

The variables in Eq. 3.12 make it very applicable to membrane whose separation mechanism depends on solution–diffusion or where affinity dominates over pore size. This means that Eq. 3.12 may work better for nanofiltration (NF), reverse osmosis (RO), and pervaporation (PV) membranes other than membrane processes where pore flow dominates. Equation 3.12 has most of the variables required to characterize membrane separation performance; however, the equation is directly independent of the membrane material solubility parameter δ_m. This is because the resultant solubility parameter difference between the solvent (i)–solute (j)–membrane (m) nullifies the effects of membrane polymer solubility parameter δ_m. For a binary system, the solvent flux through a solution–diffusion membrane permeated with a pure solvent is dependent on $(\delta_i - \delta_m)$ while $(\delta_j - \delta_m)$ may controls the solute permeability through the membrane.

Equation 3.13 may explain why the resultant difference in solubility parameter is independent on the membrane solubility parameter δ_m for a membrane permeated with a feed solution (made of solvent and solute) as shown in Eq. 3.12.

$$(\delta_i - \delta_m) - (\delta_j - \delta_m) = \delta_i - \delta_j \qquad (3.13)$$

Equation 3.12 shows some of the various parameters that affect membrane performance; however, depending on a system, design of experiment (DOE) could be employed to rule out some of the less sensitive and coupled parameters to optimize the a given separation system.

From Eq. 3.12, β can be defined as in Eq. 3.14 as a dimensionless parameter characteristic of the separation system.

$$\beta = \frac{6 \times 10^6 C_{pi}(1 - 2v)\left(\delta_i - \delta_j\right)^2}{E\rho_i K_d} = \ln\left[\left(\frac{L_f}{L_c}\right)\left(\frac{1}{1+\alpha}\right)\right] \qquad (3.14)$$

$$\Delta\delta = \left|\left(\delta_i - \delta_j\right)\right| \neq 0, \quad v \neq 1/2$$

β is a function of the membrane swelling and compaction, solvent retained in membrane, and the partial molar volume of the solvent and the membrane polymer. β is expected to a constant for a membrane separation system.

Membrane performance is defined by flux and observed rejection (R_i). The flux being how much fluid is transported across the membrane thickness to the permeate side under a driving force. Rejection is a measure of how much of the unwanted components in the bulk fluid (feed) not transported across the membrane thickness or retained by the membrane under a driving force.

Hence, increased performance refers to acceptable flux and rejection. In testing the model against experimental data, only rejection will be considered here since the present state of the model predicts only C_{pi} which related to rejection as in Eq. 2.3, i.e., $(1 - C_{pi}/C_b) \times 100\%$ (see Chap. 2, Sect. 2.2).

From Eq. 3.12, C_{pi} increases with increase in solvent density (ρ_i), Young's modulus (E), K_d, and the ratio of swelling to compaction (L_f/L_c), while C_{pi} decreases with increase in membrane extent of clamping (v) and difference between the solvent and solute solubility parameter ($\Delta\delta$). However, C_{pi} is insensitive to $\alpha = \frac{\overline{V_o}\lambda_f}{\overline{V_m}}$, this is because $\alpha <<< 1$, i.e., $\overline{V_m} >>>> \overline{V_o}$.

Hence $\frac{1}{1+\alpha} \rightarrow 1$ in Eq. 3.12.

3.2 Sensitivity of Effects of Simultaneous Swelling and Compaction on Membrane Performance

The basic functions and roles of membranes include selective transport, controlled delivery, discrimination, and separation. Examples of membranes in controlled delivery application include controlled drug delivery such as patches, e.g., nicotine patches; some oral or anal drugs and in selective transport role include contact lenses, artificial kidney, artificial liver, artificial pancreas, artificial lung (oxygenators), dialysis, and hemodialysis. In a discriminatory role include batteries, fuel cells, and sensors especially used in the detection of chemical and biological agents. Membrane in a separation role includes water purification and waste water treatments, seawater desalination and brackish water purification, solvent recovery and purification, and removal of viruses, bacteria, fungi, and some unwanted microbes [4].

With all these membrane functions, there exist a driving force or a gradient behind each basic function for transport across membranes and include pressure, concentration, partial pressure, osmotic pressure, electric field, magnetic field, and temperature.

Equation 3.12 (see Sect. 3.1) shows the term below in Eq. 3.15 as part of the relation which considers swelling and compaction in addition to the properties of the solvent and membrane. Note that swelling and compaction are in opposite direction but the magnitude not the direction is of interest here. In real life, membrane could be swollen when in contact with the preserving fluid or while in contact with the solution or solvent mixture before permeation under a driving force. The membrane undergoing permeation is compacted simultaneously.

$$\ln \left[\left(\frac{L_f}{L_c} \right) \left(\frac{1}{1+\alpha} \right) \right] \tag{3.15}$$

But $\frac{1}{1+\alpha} \rightarrow 1$ since $\overline{V_m} >>>> \overline{V_o}$ Eq. 3.15 becomes Eq. 3.16.

$$\ln \left[\left(\frac{L_f}{L_c} \right) \right] \tag{3.16}$$

But in Eq. 2.3 (see Chap. 2, Sect. 2.2) i.e.,

$$\text{Observed Rejection } (R_i) \% = \left(1 - \frac{C_{pi}}{C_b} \right) \times 100 \%$$

where C_{pi} is the permeate solute concentration and C_b the bulk solute concentration in the feed. Depending on the application or driving force swelling and/or compaction will be present or not. In some cases, swelling will dominate compaction or vice versa or swelling could be balance by compaction.

Based on the general membrane functions, several scenarios can be presented to test the model applicability to the different environments. The following cases could be a general representation to describe different polymeric membrane used in different applications. Irrespective of the driving force to Eq. 3.16 is tested to determine where it applies;

Case 1 Membrane system with both no swelling and compaction, i.e., $L_c = 0$ and $L_f = 0$. In this case, Eq. 3.16 is undefined hence the model is not applicable to this situation. This may rarely occur in real world since some swelling, compaction, or the combination may exist under the driving force however small.

Case 2 Membrane swelling with no compaction, i.e., $L_c = 0$. In this case, Eq. 3.16 is undefined hence the model is not applicable to this condition. This could mean once the membrane is swollen under the driving force some compaction should exist.

Case 3 Membrane compaction with no swelling, i.e., $L_f = 0$. In this case, Eq. 3.16 is infinity which means C_{pi} (permeate solute concentration) is infinity; however, the greatest value of C_{pi} in any separation is C_b. Infinity here is a large number such that $C_{pi} \approx C_b$ hence at $C_{pi} \approx C_b$ for a membrane system corresponds to a rejection (R_i) of 0 %. This means some swelling is necessary to improve on the rejection.

Case 4 Similar extent of swelling and compaction, i.e., $L_f \approx L_c$. In this case, Eq. 3.16 is 0 which means C_{pi} is 0 which makes the rejection 100 %. This could mean a balance between swelling and compaction is required to have good rejection. Also, it could mean that a swollen membrane needs to be compacted by the same extent to have a good rejection.

Case 5 High membrane extent of swelling with very low extent of compaction, i.e., $L_f >>>>> L_c$ such that $L_c \rightarrow 0$. Hence, Eq. 3.16 becomes

$$\underset{L_c \rightarrow 0}{\text{Lim}} \ln \left[\left(\frac{L_f}{L_c} \right) \right] \rightarrow \infty$$

which means $C_{pi} \rightarrow \infty$ (i.e., $C_{pi} \rightarrow C_b$) hence rejection $\rightarrow 0$ %. This could mean that too much swelling with little compaction lead to lowered rejection.

Case 6 Low membrane extent of swelling with high extent of compaction, i.e., $L_f <<<<<< L_c$ such that $L_f \rightarrow 0$

$$\underset{L_f \rightarrow 0}{\text{Lim}} \ln \left[\left(\frac{L_f}{L_c} \right) \right] \rightarrow 0$$

Which means $C_{pi} \rightarrow 0$ (i.e., $C_b >>>>>> C_{pi}$) hence rejection $\rightarrow 100$ %. This could mean that compaction is needed to reduce swelling will be necessary to achieve a high rejection. In general case, swelling opens up pores or membrane network and reduces rejection.

The above shows that some minimum compaction needs to be present for the model to be applicable which compaction is prevalent in most pressure-driven solution–diffusion membranes. In addition, a minimum combination is necessary.

A special case is considering pressure-driven polymeric solvent resistant membranes which undergoes both swelling and compaction. Some of these membranes are received stored in preserving solvents which will required preconditioning to remove the preserving solvent(s) before putting membrane in use. Both the preserving solvent and preconditioning process lead to the membrane sustaining some swelling before put in use. Moreover, a membrane compacted as results of permeation regains part of the compacted thickness after the applied or driving force is removed and swells again when in contact with the fluid. Generally, membrane under permeation could be constrained and subjected to the transmembrane pressure.

Let Fig. 3.2 represents a situation where a constrained membrane is swollen and compacted.

Let L_s be the total membrane thickness after a dry membrane of thickness L constrained, swells by L_f along the thickness, and ready to be permeated at a transmembrane pressure ΔP, then L_s can be represented as in Eq. 3.17.

Fig. 3.2 Schematic of a
membrane showing swelling
and compaction

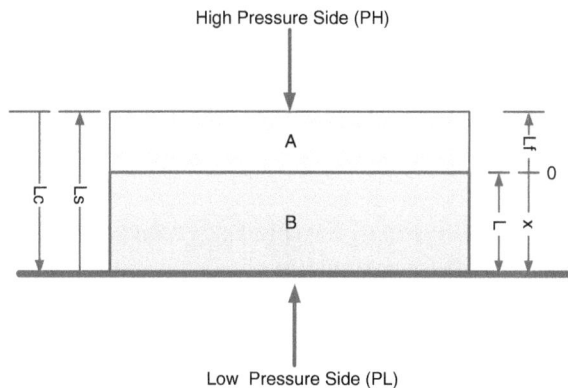

$$L_s = L + L_f \tag{3.17}$$

Generally, for pressure-driven solution–diffusion membrane, ΔP is such that the membrane compaction L_c from L_s is in the region of B from A as shown in Fig. 3.2 and can be represented as in Eq. 3.18.

$$L_c = L_f + x \quad 0 < x < L \tag{3.18}$$

where x is the distance along the dry membrane thickness L. Hence, the definition of the ratio of swelling L_f to compaction L_c can be represented as in Eq. 3.19.

$$\frac{L_f}{L_c} = \frac{L_f}{L_f + x} = \frac{1}{1 + x/L_f} \tag{3.19}$$

Case 1 For $x >>>> L_f$; $x/L_f \rightarrow \infty$ in Eq. 3.19, hence Eq. 3.16 $\rightarrow \infty$, $C_{pi} \rightarrow \infty$ then rejection $\rightarrow$ 0 %.

Case 2 For $x <<<< L_f$; $x/L_f \rightarrow 0$ in Eq. 3.19, hence Eq. 3.16 $\rightarrow 0$, $C_{pi} \rightarrow 0$ then rejection $\rightarrow$ 100 %.

Case 3 If ΔP is such that compaction in region A only, then the compaction L_c can be related to swelling L_f as in Eq. 3.20.

$$L_c = \theta L_f \quad 0 < \theta < 1 \tag{3.20}$$

where θ is a fraction showing compacted membrane thickness to swollen thickness in region A only of Fig. 3.2. The ratio of swelling L_f to compaction L_c can be represented as in Eq. 3.21.

$$\frac{L_f}{L_c} = \frac{L_f}{\theta L_f} = \frac{1}{\theta} \tag{3.21}$$

When $\theta \rightarrow 0$; $L_f/L_c \rightarrow \infty$ in Eq. 3.21, hence Eq. 3.16 $\rightarrow \infty$, $C_{pi} \rightarrow \infty$ then rejection $\rightarrow$ 0 %.

When $\theta \to 1$; $L_f/L_c \to 1$ in Eq. 3.21, hence Eq. 3.16 $\to 0$, $C_{pi} \to 0$ then rejection $\to 100 \%$.

Also, the above shows that some compaction is necessary for the model to be applicable. The above shows that the model applies for compactions in regions A and B.

3.3 Behavior of Swollen Membrane Under Permeation with Compaction

Table 3.1 is generated for explanation purposes; however, the trend is similar to what was obtained from experiment using Ultrasonic Time-domain Reflectometry (UTDR) [5] when a constrained polymeric membrane in contact with an organic feed solution is allowed to swell before compacted while permeated in a cross-flow membrane cell (see Fig. 3.3). Details of the UTDR for this application to membrane swelling and compaction is presented elsewhere by Anim-Mensah et al. [5]. The arrangement in Fig. 3.3 allows for studying the various effects of swelling, and transmembrane pressure on compaction and in situ compressive Young's modulus E on membrane performance for any system of interest while membrane is being permeated.

The arrangement above shows that while the membrane is permeated under a transmembrane pressure, it is subjected to compressive stresses and strains. Figure 3.4 shows the plots of compressive stresses (same as transmembrane pressure) versus the compressive strain for the different membrane's extent of swelling permeated with an organic solution. From the plots, the various compressive Young's moduli can be calculated. The arrangement in Fig. 3.3 permits taken into consideration the solvent interaction effect on mechanical properties. Note here that, most Young's modulus published data are from tensile test which excludes solvent interaction. Pressure-driven polymeric membranes which have free volumes and/or porosity undergoing permeation and interacting with solvents and solutes compresses and densifies depending on the applied transmembrane pressure and extent of swelling before compaction.

Let the initial thickness of a dry polymeric membrane be 150 μm, and let L_s be the final membrane thickness after swelling then the swollen membrane thickness

Table 3.1 Generated data for plotting typical membrane compressive stress–strain graphs

Transmembrane pressure (MPa)	0 % swelling		20 % swelling		40 % swelling		60 % swelling		80 % swelling		100 % swelling	
	L_s	L_c	L_s	L_c	L_s	L_c	L_s	L_c	L_s	L_c	L_s	L_c
0	150	0	180	0	210	0	240	0	270	0	300	0
0.69	150	30	180	60	210	130	240	180	270	210	300	240
1.03	150	45	180	85	210	155	240	195	270	230	300	260
1.38	150	55	180	100	210	170	240	220	270	250	300	284
1.72	150	60	180	115	210	190	240	230	270	262	300	290

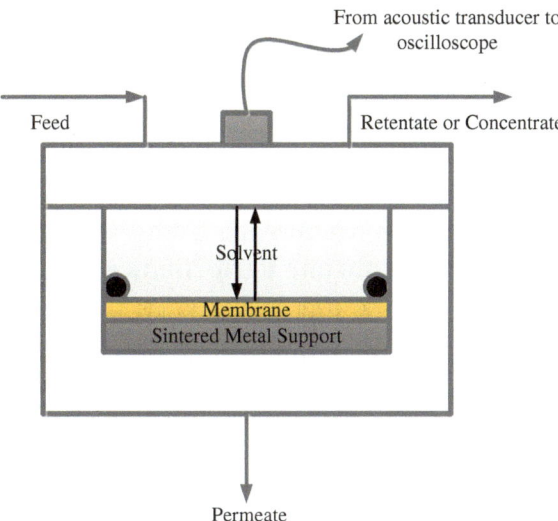

Fig. 3.3 A cross-flow membrane cell accommodating UTDR transducer for real-time membrane swelling and compaction simultaneous with permeation analysis [5]

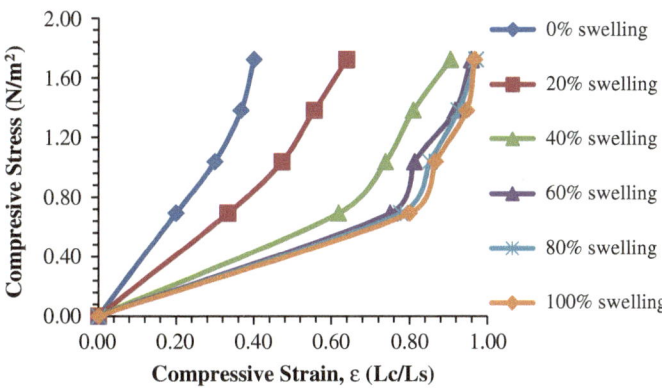

Fig. 3.4 Typical stress–strain plot of membranes compacted while permeated

L_f can be calculated as $L_f = L_s - 150$ μm. Let the swollen membrane thickness L_s be compacted while permeated under a transmembrane pressure (ΔP) or compressive stress (σ) be L_c then the membrane compressive strain (ε) can be calculated as L_c/L_s. All swelling and compaction dimensions are in microns (μm). Table 3.1 presents a typical result generated for explanation purposes when for an initial dry membrane swollen and compacted while permeated.

Figure 3.4 shows a typical compressive stress–compressive strain plot of a membrane swollen and undergoing simultaneous compaction with permeation from Table 3.1. The graphs in Fig. 3.4 could be typical of a constrained membrane exposed to different solvents or solutions for the same length of time and compacted while permeated or a membrane exposed to a solvent or solution for different length of time compacted while permeated.

3.4 Effects of Composite Membrane Compressive Young's Modulus (*E*) on Membrane Performance

From Eq. 3.12 (see Sect. 3.1), an increase in composite membrane Young's modulus (E) increases C_{pi}. Figure 3.5 shows plots of C_{pi} or membrane performance with transmembrane pressure for the different membrane composite Young's modulus using 40 % membrane swelling, $\alpha\left(\frac{V_0\lambda_f}{V_m}\right) = 2.96\text{e-}4$, $\rho_i = 850$ kg/m^3, $\Delta\delta = 5.2$, $\nu = 1.0$ and $K_d = 0.001$. The Young's modulus will be referred to as in situ since it will varying dependent on the solvent interaction and may be different from what is published in literature. In Fig. 3.5, at each E value, the C_{pi} increased with the transmembrane pressure. In addition, increasing E values resulted in decreasing C_{pi}.

By definition, compressive Young's modulus is the ratio of the stress to the strain. A low E values mean a high strain at a low stress (transmembrane pressure). This means the membrane densifies considerably at a given transmembrane pressure reducing the membrane-free volume, increasing membrane resistance, or reducing the swollen network decreasing the flux of solute molecules across the membrane. This low E could also show the effects of possible increased membrane densification on membrane performance.

At a given E, an increase in C_{pi} with the transmembrane pressure could be explained by the fact that the solute molecule has enough driving forces to travel across the shortened diffusion path as results of membrane compaction. The effects depend on the affinity between the membrane and the solute molecules as well as the membrane pores size.

Fig. 3.5 The effects of composite membrane Young's modulus at 40 % membrane swelling on performance

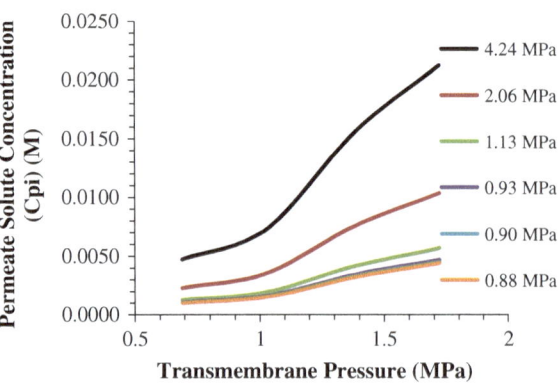

Fig. 3.6 The effects of solvent density (ρ_i) on membrane performance

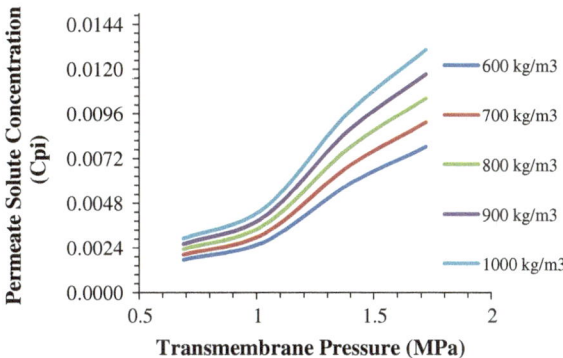

3.5 Effects of Solvent Density (ρ_i) on Membrane Performance

In Eq. 3.12 (see Sect. 3.1), holding the rest of the variables constant, it is expected that, an increase in the solvent density (ρ_i) increases C_{pi}. This means the membrane performance observed rejection decreases. This appears counter intuitive, however, surprisingly increasing solvent densities as shown in Table 2.1 (see Chap. 2, Sect. 2.1) generally follow the trend below:

$$\text{Polar Aprotic} > \text{Polar Protic} > \text{Non-Polar}$$

(see Table 2.1).

Interestingly, the general trend above is similar to aggressiveness of the organic solvents (Table 2.1) which is related to extent of membrane swelling resulting in the increased C_{pi} as the solvent densities increased.

Figure 3.6 shows a plot of the various solvent densities on membrane performance using 40 % membrane swelling, $\alpha\left(\frac{\overline{V_0}\lambda_f}{\overline{V}_m}\right) = 2.96\text{e-}4$, $\Delta\delta = 5.2$, $\nu = 1.0$, $E = 2.21\,\text{MPa}$ and $K_d = 0.001$. From the plot, the C_{pi} generally increased from nonpolar solvent through to polar aprotic as expected and explained by the aggressiveness trends of the solvents.

3.6 Effects of Solubility Parameter Differences Between Solvents and Solutes ($\Delta\delta$) on Membrane Performance

Equation 3.12 (see Sect. 3.1) shows the sensitivity of C_{pi} to $\Delta\delta$, i.e., ($\delta_i - \delta_j$). Interestingly, it shows that the performance is independent of membrane solubility parameter δ_m. An increase in $\Delta\delta$ is expected to decrease C_{pi} which means an increase membrane performance. Though, the total solubility parameter will not always result in a good prediction, it thus provides a guide to select good separation systems. In some cases, one of the individual contributions to the total solubility parameter, i.e., dispersion, hydrogen, or polar contribution could dominate the overall solubility parameter.

From Eq. 3.12, it could generally be deduced that for a solution–diffusion membrane separation system;

- A high affinity between a solvent and a solute means δ_i and δ_j are very close; if the solvent likes the membrane, then the solute will also like the membrane. This could result in high flux and low solute rejection. The extent of affinity between the membrane and solute will determine the formation of the concentration polarization layer as well as irreversible fouling.
- A high affinity between solvent and solute means δ_i and δ_j are very close; if the solvent does not like the membrane, then the solute may also not also like the membrane. This could results in low flux and low solute rejection.
- A low affinity between solvent and solute means δ_i and δ_j are not very close; if the solvent like the membrane then the solute may not. This could results in high flux with high solute rejection. This is provided that affinity between the membrane and solvent is such that it will not result in excessive membrane swelling or even membrane dissolution.
- A low affinity between solvent and solute means δ_i and δ_j are not very close; if the solvent does like the membrane then the solute could like the membrane. If the solute likes the membrane, this could results in low flux and low solute rejection. The solute having affinity for the membrane could result in fast solute buildup and possible irreversible fouling in the long term.

Figure 3.7 shows a plot of the solubility parameter difference between the solvent and solute on membrane performance using 40 % membrane swelling, $\alpha\left(\frac{V_0 \lambda_f}{V_m}\right) = 2.96e\text{-}4$, $\nu = 1.0$, $E = 2.21$ MPa, $\rho = 850$ kg/m^3 and $K_d = 0.001$. In Fig. 3.7, all the plots showed C_{pi} increasing with decrease in $\Delta\delta$ at a given transmembrane pressure. In addition to a given $\Delta\delta$ value, the C_{pi} increased with the transmembrane pressure.

Fig. 3.7 The effects of solute and solvent solubility parameter difference ($\Delta\delta$) on performance

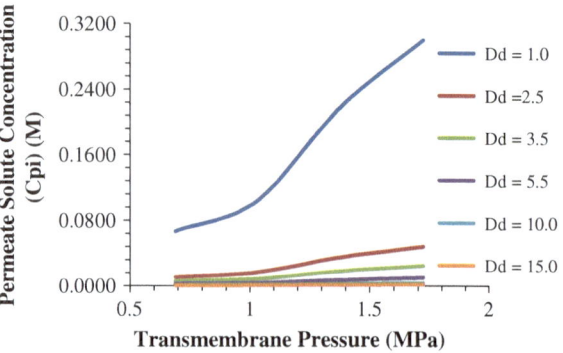

Generally, small $\Delta\delta$ value depicts closeness in the cohesive energies of the solvent and solute molecules, i.e., high affinity [6] which is expected to result in high C_{pi} values and vice versa. This means closeness in cohesive energies between the solvent and solute may lead to low membrane performance. At a given $\Delta\delta$, an increasing C_{pi} with the transmembrane pressure could be explained by the fact that the solute molecule has enough driving force to travel across the membrane. Also, it could be explained by compaction associated with the compacted membrane thickness shortening the diffusion length hence C_{pi} increases.

3.7 Effects of Poisson Ratio (v) on Membrane Performance

Poisson ratio (v) is defined as the ratio of the transverse contraction strain to the longitudinal extension strain for a material under a stretching or compression force [7]. However, most of the published data in Table 2.2 (see Chap. 2, Sect. 2.5) are obtained from elastic and free unconstrained material surfaces under tension or compression. Generally, for objects with free unconstraint surfaces under tension or compression, $-1 < v < 0.5$ is true for stability. For a membrane which could be porous, constrained undergoing simultaneous compression with permeation could be deformed inelastically and undergo densification; Poisson ratio (v) could be outside the normal range. This means v could be outside the range of $-1 < v < 0.5$, i.e., $v < -1$ and $v > 0.5$ and still be stable [8].

Generally, for a constrained swollen membrane compacted while under permeation, the transverse strain is higher than the longitudinal strain, in addition to the membrane could inelastically be deformed while it densifies making the Poisson ratio (v) approach large values outside the normal range. Generally, pressure-driven membranes are constrained and compressed leading to densification while in operation hence can be described with Poisson ratio outside the normal range. In addition, different membrane materials with different solvent interactions during permeation could result in different v for the same applied transmembrane pressure hence behave differently when constrained and permeated.

Figure 3.8 shows a plot of the effect of Poisson ratio (v) on membrane performance using 40 % membrane swelling, $\alpha\left(\frac{V_o\lambda_f}{V_m}\right) = 2.96\text{e-}4$, $v = 1.0, \Delta\delta = 5.2, E = 2.21\,\text{MPa}, \rho = 850\,\text{kg/m}^3$ and $K_d = 0.001$. Figure 3.8 shows that for all v values, C_{pi} increased with the transmembrane pressure. At a given transmembrane pressure, an increase in Poisson ratio (v) decreases the C_{pi} which means the membrane performance increased. Within the value of Poisson ratio (v) used in Fig. 3.8, it is apparent that clamping or constraining membrane sufficiently could lead to better performance. This is because of sufficient membrane densification.

The extent of densification is dependent on the extent of constraint, applied stress, membrane material type, and solvent interaction with the membrane. When

Fig. 3.8 The effects of different polymer Poisson ratio on performance

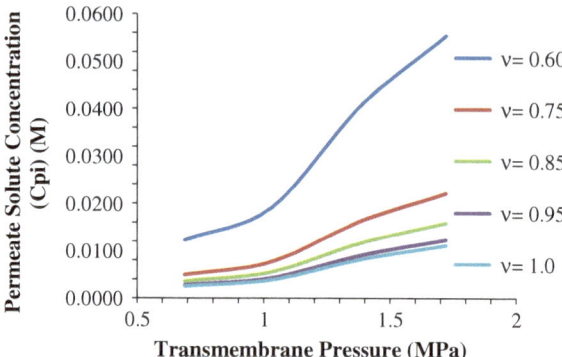

a membrane densifies, the free volume reduces, pore constricts, membrane morphology changes, and/or membrane resistance increases. This could explain why membrane constraint is necessary to achieve considerable performance as shown in Fig. 3.8. However, excessive compaction or constraint could impact membrane performance. At a given v, an increasing C_{pi} with the transmembrane pressure could be explained by the fact that the solute molecule has enough driving force to travel across the membrane.

3.8 Effects of $\alpha = \frac{\overline{V_o}\lambda_f}{\overline{V_m}}$ on Membrane Performance

From Eq. 3.12 (see Sect. 3.1) $\alpha = \frac{\overline{V_o}\lambda_f}{\overline{V_m}}$ is insensitive, this is because $\alpha \lll 1$, i.e., $\overline{V_m} \ggg \overline{V_o}$ or $\frac{1}{1+\alpha} \to 1$; hence, with the rest of the parameters kept constant, a change in α is expected not to cause a change on membrane performance. Figure 3.9 shows the effect of $\alpha = \frac{\overline{V_o}\lambda_f}{\overline{V_m}}$ on membrane performance

Fig. 3.9 Effects of varying α on membrane performance (C_{pi})

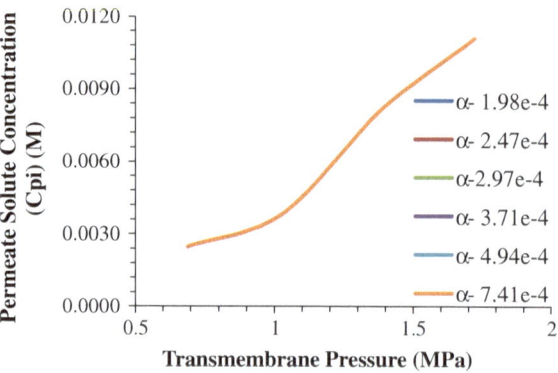

at the different transmembrane pressure using 40 % membrane swelling, $v = 1.0$, $\Delta\delta = 5.2$, $E = 2.21$ MPa, $\rho = 850$ kg/m^3 and $K_d = 0.001$. At any given transmembrane pressure, the C_{pi} is the same for all α values which means α effect is minimal within the values considered here.

Excessive compaction, as shown by increased transmembrane pressure, leads to poor membrane performance.

3.9 Effects of Swelling on Membrane Performance

Figure 3.10 shows the effect of membrane swelling on performance for the different transmembrane pressure using $\alpha\left(\frac{V_0\lambda_f}{V_m}\right) = 2.96e\text{-}4$, $v = 1.0$, $\Delta\delta = 5.2$, $E = 2.21$ MPa, $\rho = 850$ kg/m^3, and $K_d = 0.001$.

From Fig. 3.10, it is apparent that the extent of swelling considered here did not have significant effect on membrane performance. This is because the swelling considered here was not extreme to result in extreme densification that will lead to lowered membrane performance through excessive compaction or membrane network opening. However, a balance between swelling and compaction is necessary to achieve good membrane performance. Excessive compaction, as shown by increased transmembrane pressure, leads to poor membrane performance.

Fig. 3.10 Effects of membrane swelling (%) on performance (C_{pi}) at Poisson ratio (v) of 1

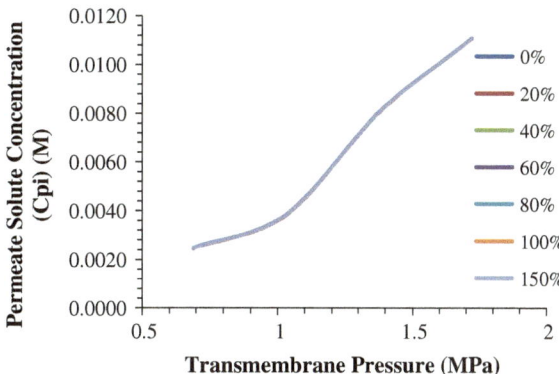

3.10 Effects of Membrane Swelling and Compressive Young's Modulus on Membrane Performance

Here, the swelling and compaction associated with each Young's modulus (E) obtained from Table 3.1 (see Sect. 3.3) and Fig. 3.4 are used in addition to $\alpha\left(\frac{\overline{V_0}\lambda_f}{V_m}\right) = 2.96\text{e-}4, \nu = 1.0, \Delta\delta = 5.2, \rho = 850 \text{ kg/m}^3$ and $K_d = 0.001$ to determine the effect of both membrane swelling and in situ E on membrane performance. Figure 3.11 shows the plots of the effects of varying swelling and E on membrane performance with transmembrane pressure.

The above plots show that the lowest swelling (1 %) and highest E (4.24 MPa) results in the lowest membrane performance compared with the highest swelling (100 %) and lowest E (0.87 MPa). This shows that swelling associated with low E could lead to densification which results in high membrane performance. However, there is a limit to extent of swelling at which performance is high. The low swelling and high E could be due to the fact that the solvent and membrane have little affinity which the low swelling could lead to high E. At low swelling and the transmembrane pressure, the membrane does not densify enough to reduce the free volume to restricted solute transport across the membrane. This shows that reasonable swelling or affinity between the solvent and membrane is required for improved solution–diffusion membrane performance.

Comparing Fig. 3.11 (Sect. 3.10) and Fig. 3.5 (Sect. 3.4), it could be concluded that in situ low E dominates swelling; however, low E could be as results of the membrane material and interaction with the solvent as well as extent of swelling. Hence some reasonable amount of swelling is required for improved membrane performance.

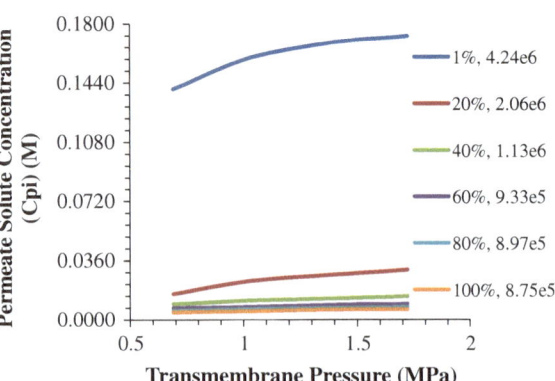

Fig. 3.11 Effect of membrane swelling and compressive Young's modulus (E) on membrane performance (C_{pi})

3.11   Effects of Solute Distribution Coefficient (K_d) on Membrane Performance

The variation of K_d (i.e., $C_{pi}/C_m \leq 1$) with transmembrane pressure, temperature, flux, and feed concentration for a given separation system is unknown; however, it could be extracted given an experimental data by using Eqs. 2.1 and 2.2 (see Chap. 2, Sect. 2.2). Here various K_d variation with the transmembrane pressure trends, i.e., constant, increasing, decreasing, increasing and decreasing or decreasing and increasing are assumed to determine their effects on membrane performance. Table 3.2 shows K_d values and trends at the various transmembrane pressures used to test the model in addition to using 40 % membrane swelling, $\alpha\left(\frac{V_o \lambda_f}{V_m}\right) = 2.96\text{e-}4, v = 1.0, \Delta\delta = 5.2, E = 2.21$ MPa, and $\rho = 850$ kg/m^3.

Figure 3.12 shows a plot of C_{pi} variation with transmembrane pressure for the various K_d trends.

In Fig. 3.12, the increasing K_d trend with the transmembrane pressure resulted in a remarkable increasing C_{pi} trend with the transmembrane pressure while the constant resulted in a slightly increasing trend. The decreasing K_d trend resulted in a slightly increasing and decreasing C_{pi} trend. The increasing and deceasing K_d trend resulted in a remarkable increasing and decreasing C_{pi} trend with the

Table 3.2   Assumed K_d values used to test the model's behavior on membrane performance trend

Transmembrane pressure (MPa)	Constant K_d values	Increasing K_d values trend	Decreasing K_d values trend	Increasing and decreasing K_d values trend	Decreasing and increasing K_d values trend
0.69	0.001	0.001	0.004	0.001	0.004
1.03	0.001	0.002	0.003	0.004	0.001
1.38	0.001	0.003	0.002	0.004	0.001
1.72	0.001	0.004	0.001	0.001	0.004

Fig. 3.12   The effects of K_d variation with transmembrane pressure on membrane on performance

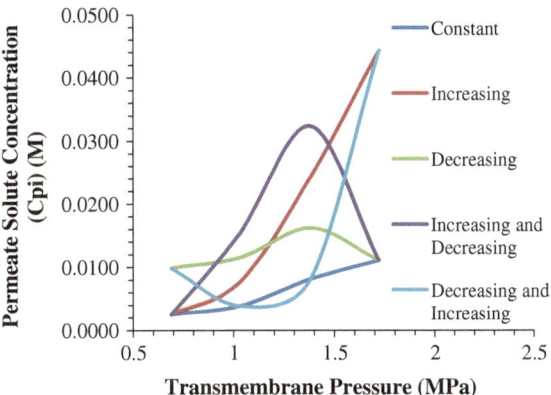

transmembrane pressure while the decreasing and increasing K_d trend also resulted in a remarkable decreasing and increasing C_{pi} trend with the transmembrane pressure. The above shows that within the values of K_d used, the decreasing and increasing, the decreasing K_d trends did not result in remarkable effects on membrane performance.

References

1. Weber, A. Z., & Newman, J. (2004). A theoretical study of membrane constraint in polymer-electrolyte fuel cells materials, interfaces, and electrochemical phenomena. *AIChE Journal, 50*, 3215–3226.
2. Timoshenko, S. (1958). *Strength of materials, part II, advanced theory and problems* (3rd ed.). Toronto, Melbourne: Van Nostrand Reihnold Company Ltd.
3. Orwoll, R. A., & Arnold, P. A. (1996). Polymer–solvent interaction parameter χ. In J. E. Mark (Ed.), *Physical properties of polymers handbook*. New York: AIP Press.
4. Anim-Mensah, A. R., Franklin, J. E., Palsule, A. S., Salazar, L. A., & Widenhous, C. W. (2010). Characterization of a biomedical grade silica-filled silicone elastomer using ultrasound, advances in silicones and silicone-modified materials. *ACS Symposium Series, 1051*, 85–98. (Chap. 8).
5. Anim-Mensah, A. R. (2007). Evaluation of solvent resistant nanofiltration (SRNF) membranes for small-molecule purification and recovery of polar aprotic solvents for re-use (Ph.D. thesis, University of Cincinnati, OH, 2007).
6. Anim-Mensah, A. R., Mark, J. E., & Krantz, W. B. (2007). Use of solubility parameters for predicting the separation characteristics of poly(dimethylsiloxane) and siloxane-containing membranes, science and technology of silicones and silicone-modified materials. *ACS Symposium Series, 964*, 203–219. (Chap. 14).
7. Fung, Y. C. (1968). *Foundation of solid mechanics*. Englewood, NJ: Prentice-Hall.
8. Greaves, G. N., Greer, A. L., Lakes, R. S., & Rouxel, T. (2011). Poisson's ratio and modern materials. *Nature Materials, 10*, 823–837.

Chapter 4
Developed Model Application to Aqueous— Organic and Purely Organic Separation and Purification Systems

Abstract Reduced versions of the developed model are used to predict published membrane separation and purification trends of three organic systems involving polar and polar aprotic solvents. The model predicted the trends and presented the key parameters involved in the pervaporation of aqueous-ethanol solutions using unfluorinated and fluorinated polysiloxane-imide (PSI) membranes [1] as well as pervaporation of methyl tertiary butyl ether (MTBE) and Butyl Acetate (BuAC) aqueous solutions using siloxane-urethane (SU) membranes [2]. Moreover, the model predicted the trends and presented key parameters of a polyimide (PI) nano-filtration membrane separation and purification processes involving organic solute (leucine) in the following organic solvents and mixtures; 1-butanol, 1-butanol and dimethylformamide (DMF), and 1-butanol and N-methylpyrrolidone (NMP) [3]. Ethanol and 1-butanol are polar protic organic solvent while MTBE, BuAC, DMF and NMP are polar aprotic organic solvents.

Keywords Model   application · Ethanol   dehydration,   Separation · Purification · Organic solvents · Membranes · Fluorinated membranes · MTBE separation · BuAC separation · Solvent-resistant · Polymeric   membranes · Polysiloxaneimide   (PSI) · Dimethylformamide (DMF) separation · Pervaporation · Nanofiltration · Polyimide

In this chapter, the model developed in Eq. 3.12 (see Chap. 3, Sect. 3.1) is used to predict the separation and purification organic solvent trends of research works of Krea et al. [1], Czerwiński et al. [2], and Anim-Mensah [3] sequentially in the Sects. 4.1, 4.2, and 4.3, respectively.

Krea et al. research work involved the use of unfluorinated and fluorinated polysiloxane-imide (PSI) pervaporation membranes for the separation and purification of ethanol from aqueous ethanol solutions. Ethanol is a polar protic organic solvent. Moreover, Czerwiński et al. research work involved the use of siloxane–urethane (SU) pervaporation membranes for the separation and purification of methyl tertiary butyl ether (MTBE) and BuAC from their aqueous solutions. Both MTBE and BuAC are polar aprotic organic solvents. Finally, Anim-Mensah [3]

© The Author(s) 2015
A. Anim-Mensah and R. Govind, *Prediction of Polymeric Membrane Separation and Purification Performances*, SpringerBriefs in Molecular Science,
DOI 10.1007/978-3-319-12409-4_4

research work involved the use of polyimide nanofiltration membranes for the separation and purification of leucine (an organic solute) from organic solvent and its mixtures comprised of 1-butanol, dimethylformamide (DMF), and N-methylpyrrolidone (NMP). 1-Butanol is a polar protic organic solvent, while both DMF and NMP are polar aprotic solvents. Here, the membrane is serving to purifying the leucine as well as recovering the organic solvents for reuse.

Most of the variables in Eq. 3.12 (see Chap. 3, Sect. 3.1) are obtained from experiments since published data are not available; however, the solvent (δ_i) and solute (δ_j) solubility parameters, solvent density (ρ_i), molar volume of membrane polymer ($\overline{V_m}$), and solvent ($\overline{V_o}$) are available in published data; for this reason, reduced versions of the model will be used for the trend prediction.

For lack of information, approximations from Eq. 3.12 will be used as a quick way to predict membrane performance trend. For illustration purposes in each of the following situations when the rest of the parameters are kept constant in Eq. 3.12 (see Chap. 3, Sect. 3.1); C_{pi} is proportional to solvent densities (ρ_i); C_{pi} is proportional to $\rho_i/(\delta_i - \delta_j)^2$, i.e., $\rho/\Delta\delta^2$; and C_{pi} is proportional to $\psi = \frac{10^{-6}E\rho}{6(\delta_i - \delta_j)^2}\ln\left[\left(\frac{L_t}{L_c}\right)\left(\frac{1}{1+\alpha}\right)\right]$ are used to predicted C_{pi} trends. Note that an increase in C_{pi} means a decrease in membrane performance, i.e., the observed rejection (R_i) because of the relation between R_i and C_{pi}, i.e., $R_i = (1 - C_{pi}/C_b) \times 100$ % (see Chap. 2, Sect. 2.2). A plot of experimental R_i versus approximated C_{pi}, i.e., (ρ_i), $\rho/\Delta\delta^2$ and ψ are expected to be graph with a negative trend. However, a plot of experimental C_{pi} versus approximated C_{pi}, i.e., (ρ_i), $\rho/\Delta\delta^2$ and ψ are expected to be graph with a positive trend.

4.1 Dehydration of Aqueous Ethanol Using Unfluorinated and Fluorinated Polysiloxane-Imide (PSI) Pervaporation Membranes with Varying Siloxane wt%

Here, the model is used to predict the dehydration trends of 10 and 50 wt% of ethanol solution using pervaporation through unfluorinated and fluorinated PSI membranes with varying wt% of siloxane at 40 °C [1]. Here, since water is unwanted in the permeate and requires to be rejected by the membrane, it will be used as the reference in order to use the definition of rejection in this context for pervaporation. Figure 4.1 shows the concentration of ethanol in the permeate using unfluorinated PSI membrane with varying siloxane concentration permeated with 10 and 50 wt% of ethanol solutions.

Figure 4.2 shows the concentration of ethanol in the permeate using fluorinated PSI membrane with varying siloxane concentration permeated with 10 and 50 wt% of ethanol solutions.

Table 4.1 shows the properties of the individual solvent and the solvent mixtures.

Tables 4.2, 4.3, 4.4, 4.5, 4.6, and 4.7 are generated from Figs. 4.1 and 4.2, respectively, with reference to water to be rejected by the membranes.

Fig. 4.1 Concentrations of ethanol in permeate using unfluorinated PSI with varying wt% siloxane in membrane [1]

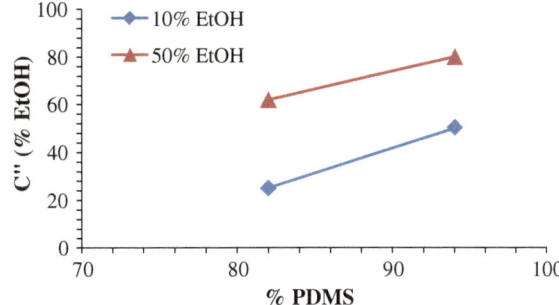

Fig. 4.2 Concentration of ethanol in permeate using fluorinated PSI with varying wt% siloxane in membrane [1]

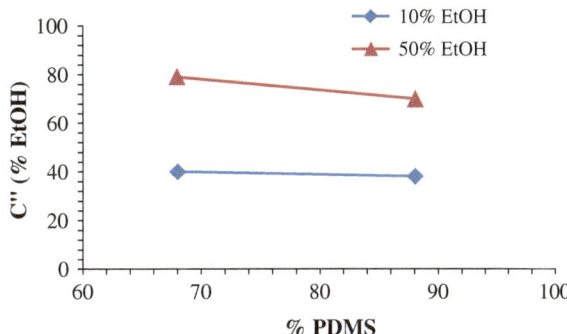

Table 4.1 Properties of the individual solvent and solvent mixtures

Solvents	Density (g/ml)	Solubility parameter $(MPa^{0.5})$
Water	1.000	47.9
Ethanol	0.789	26.0
10 wt% ethanol–water solution	0.979	45.71
50 wt% ethanol–water solution	0.895	36.95

Table 4.2 Summary from Figs. 4.1 and 4.2 [1]

	PSI membrane		FPSI membrane	
	82 wt% siloxane in membrane	94 wt% siloxane in membrane	68 wt% siloxane in membrane	88 wt% siloxane in membrane
wt% ethanol in feed solution	wt% ethanol in permeate			
10 (i.e., 90 water)	25 (i.e., 75 water)	50 (i.e., 50 water)	40 (i.e., 60 water)	38 (i.e., 62 water)
50 (i.e., 50 water)	62 (i.e., 38 water)	80 (i.e., 20 water)	80 (i.e., 20 water)	70 (i.e., 30 water)

Table 4.3 Pervaporation of 10 and 50 wt% ethanol feed solution using unfluorinated polysiloxane-imide (PSI) membrane

	82 wt% siloxane in membrane		94 wt% siloxane in membrane	
wt% water in feed (C_b)	90	50	90	50
Feed density (g/ml)	0.9787	0.8945	0.9787	0.8945
wt% water in permeate (C_p)	75	38	50	20
Actual rejection of water by membrane (R)	16.7	24	44.4	60
Predicted membrane performance as $\rho/\Delta\delta^2$	2.14×10^3	1.95×10^3	2.14×10^3	1.95×10^3

Table 4.4 Pervaporation of 10 and 50 wt% ethanol feed solution using fluorinated polysiloxane-imide (FPSI) membranes

	68 wt% siloxane in membrane		88 wt% siloxane in membrane	
wt% water in feed (C_b)	90	50	90	50
Feed density (g/ml)	0.9787	0.8945	0.9787	0.8945
wt% water in permeate (C_p)	60	20	62	30
Actual observed rejection of water by membrane (R)	33.3	60	31.1	40
Predicted membrane performance as $\rho/\Delta\delta^2$	2.14×10^{-3}	1.95×10^{-3}	2.14×10^{-3}	1.95×10^{-3}

Table 4.5 Pervaporation of 10 and 50 wt% ethanol feed solution using unfluorinated polysiloxane-imide (PSI) membranes

$\rho/\Delta\delta^2 \, (\times^{-3})$	82 wt% siloxane in membrane PSI		94 wt% siloxane in membrane PSI	
	C_{pi}	R	C_{pi}	R
1.95	38.0	24.0	20.0	60.0
2.14	75.0	16.7	50.0	44.4

Table 4.6 Pervaporation of 10 and 50 wt% ethanol feed solution using fluorinated polysiloxane-imide (FPSI) membranes

$\rho/\Delta\delta^2 (\times 10^{-3})$	68 wt% siloxane in membrane FPSI		88 wt% siloxane in membrane FPSI	
	Concentration of water in permeate C_{pi} (%)	Water observed rejection by membrane (%)	Concentration of water in permeate C_{pi} (%)	Water observed rejection by membrane (%)
1.95	20.0	60.0	30.0	40.0
2.14	60.0	33.3	62.0	31.1

Table 4.7 Pervaporation of 10 and 50 wt% ethanol feed solution using unfluorinated and fluorinated polysiloxane-imide (PSI) membranes [1]

Feed solution	Feed solution density	Unfluorinated polysiloxane-imide (PSI) membranes		68 wt% siloxane in membrane FPSI	
		82 wt% siloxane in membrane PSI	94 wt% siloxane in membrane PSI	Fluorinated polysiloxane-imide (FPSI) membranes	88 wt% siloxane in membrane FPSI
		wt% water in permeate			
10 % ethanol	0.98	75	50	60	62
50 % ethanol	0.89	38	20	20	30

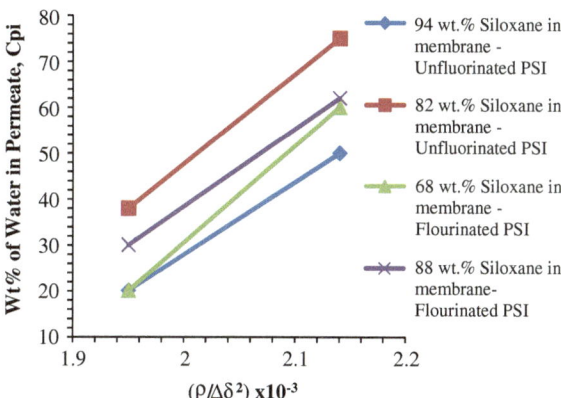

Fig. 4.3 Experimental permeate water concentration C_{pi} versus predicted permeate water concentration ($\rho/\Delta\delta^2$) showing expected direct relationship

From the above information, Table 4.3 could be derived considering C_{pi} is proportional to $\rho/\Delta\delta^2$.

Figure 4.3 shows a plot of the experimental solute (water) concentration in the permeate C_{pi} from experiment versus the calculated C_{pi}, i.e., $\rho/\Delta\delta^2$ for both the PSI and FPSI membrane with varying wt% of siloxane in the membranes.

The trend of the plots in Fig. 4.3 confirms the calculated C_{pi} values, i.e., $\rho/\Delta\delta^2$ are in accordance with the experimental trends.

Figure 4.4 shows a plot of the actual solute (water) rejection from experiment versus the calculated C_{pi}, i.e., $\rho/\Delta\delta^2$ for both the PSI and FPSI membrane with varying wt% of siloxane in the membranes.

All the actual solute rejection plots in Fig. 4.4 show a decreasing trend generally or inverse relation with $\rho/\Delta\delta^2$ confirming the calculated values' trends are in accordance with the experimental trends.

Figure 4.5 shows a plot of the effects of solvent densities on the performance of the membrane to reject water transport across them. From Eq. 3.12 (see Chap. 3, Sect. 3.1), membrane performance is affected by a general increase solvent density for organic solvents and Fig. 4.5 confirms the observation. Note that the solvent mixture density is used here (Table 4.7).

Fig. 4.4 Experimental membrane water observed rejection (R) versus predicted permeate water concentration ($\rho/\Delta\delta^2$) showing expected inverse relationship

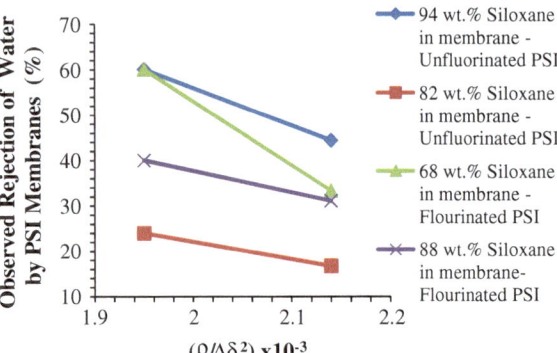

Fig. 4.5 Experimental permeate water concentration C_{pi} versus solvent densities

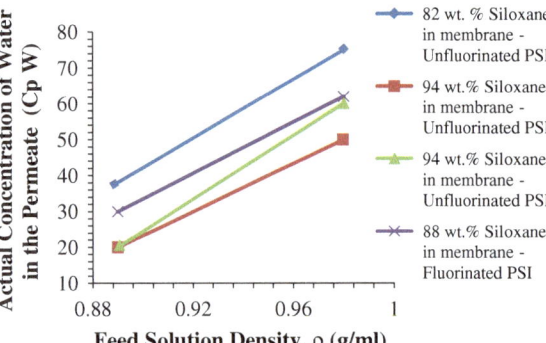

4.2 Dehydration of Aqueous Methyl Tertiary Butyl Ether (MTBE) and Butyl Acetate Using Siloxane–Urethane Pervaporation Membranes with Varying Siloxane wt%

The reduced form of Eq. 3.12 (see Chap. 3, Sect. 3.1) is used here to predict the performance of dehydration of aqueous MTBE and butyl acetate solutions since most of the variables are not presented in the paper. The dehydration involves 1.8 and 0.25 wt% aqueous feed solutions of MTBE and butyl acetate, respectively, pervaporated through siloxane–urethane membranes with varying siloxane wt% at 313 K (40 °C) [2].

Figure 4.6 shows the separation factors ($\alpha_{org/water}$) for the various SU membranes with varying siloxane wt% pervaporated with 1.8 and 0.25 wt% aqueous feed solutions of MTBE and butyl acetate, respectively [2].

Tables 4.9 and 4.10 are generated from Table 4.8 and present the properties of the solvents involved. In addition, Tables 4.9 and 4.10 present the separation factor defined as water relative to organic (i.e., $\eta_{water/org} = 1/\alpha_{org/water}$) and the calculated value of $\rho/\Delta\delta^2$ for the aqueous MTBE and BUAC system, respectively. Note that the solvent mixture density is used here.

Table 4.11 shows the information used in plotting the graphs in Figs. 4.7 and 4.8.

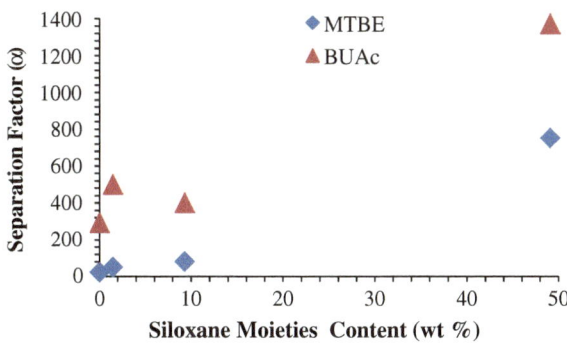

Fig. 4.6 Separation factor ($\alpha_{org/water}$) of MTBE and BuAC dehydration using siloxane-urethane (SU) membrane with varying wt% of siloxane [2]

Table 4.8 Summary of separation factor ($\alpha_{org/water}$) of MTBE and BuAC through SU membranes

wt% siloxane in membrane	0	1.46	9.3	49.05
Aqueous feed	Separation factor (α) based on organic solvent to water			
1.8 wt% MTBE (i.e., 98.2 % water)	23	50	80	750
0.25 wt% BuAC (i.e., 99.75 % water)	293	500	400	1,370

Table 4.9 Properties of the solvents and separation factors for MTBE system

wt% siloxane in membrane	0	1.46	9.3	49.05
wt% MTBE in aqueous feed solution	1.8	1.8	1.8	1.8
Density of 1.8 wt% MTBE in 98.2 % water	0.9953	0.9953	0.9953	0.9953
Separation factor ($\alpha_{org/water}$)	23	50	80	750
Separation factor ($\eta_{water/org} = 1/\alpha_{org/water}$)	4.35×10^{-2}	2.00×10^{-2}	1.25×10^{-2}	1.33×10^{-3}
Membrane performance predicted by $\rho/\Delta\delta^2$	6.07×10^{-4}	6.07×10^{-4}	6.07×10^{-4}	6.07×10^{-4}

Table 4.10 Properties of the solvents and separation factors for BUAC system

wt% siloxane in membrane	0	1.46	9.3	49.05
wt% BuAC in aqueous feed solution	0.25	0.25	0.25	0.25
0.25 wt% BuAC in water (i.e., 99.75 wt% water)	0.9997	0.9997	0.9997	0.9997
Separation factor ($\alpha_{org/water}$)	293	500	400	1,370
Separation factor ($\eta_{water/org} = 1/\alpha_{org/water}$)	3.41×10^{-3}	2.00×10^{-3}	2.50×10^{-3}	7.30×10^{-4}
Membrane performance predicted by $\rho/\Delta\delta^2$	1.08×10^{-3}	1.08×10^{-3}	1.08×10^{-3}	1.08×10^{-3}

Table 4.11 Summary of calculated solvent mixture densities and separation factors for the various SU membranes

	ρ (g/ml)	$\rho/\Delta\delta^2$ (10^{-3})	0 % siloxane $\eta_{\text{water/org}}$	1.46 % siloxane $\eta_{\text{water/org}}$	9.3 % siloxane $\eta_{\text{water/org}}$	49.05 % siloxane $\eta_{\text{water/org}}$
MTBE— water feed	0.9953	0.61	4.35×10^{-2}	2.00×10^{-2}	1.25×10^{-2}	1.33E−03
BuAC— water feed	0.9997	1.08	3.41×10^{-3}	2.00×10^{-3}	2.50×10^{-3}	7.30×10^{-4}

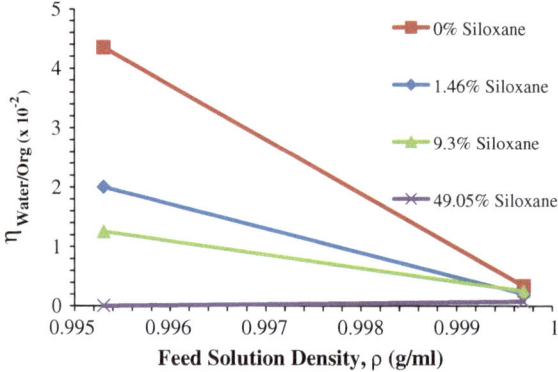

Fig. 4.7 Experimental permeate water concentration C_{pi} versus predicted permeate water concentration $(\rho/\Delta\delta^2)$ showing expected inverse relationship

Fig. 4.8 Experimental permeate solute (water) concentration C_{pi} versus predicted permeate water concentration $(\rho/\Delta\delta^2)$ showing expected inverse relationship

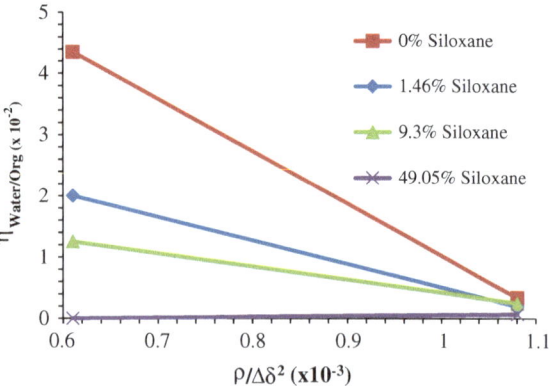

Figures 4.7 and 4.8 show the expected trend of the plots of the separation factor of water to organic, i.e., $\eta_{\text{water/org}}$ versus the solvent densities and $\eta_{\text{water/org}}$ versus the calculated membrane performance (C_{pi}, i.e., $\rho/\Delta\delta^2$), respectively.

The trends in Figs. 4.7 and 4.8 are in accordance with the model predictions.

4.3 Solute Purification and Solvents Recovery for Reuse Using Pressure-Driven Polyimide (PI) Nanofiltration (STARMEM-122) Membrane

Polyimide (PI) STARMEM-122 nanofiltration membrane is used for solute purification and solvents recovery from a solution comprised of a single solvent or solvent mixtures [3]. The feed solutions involved here are (1) leucine in 1-butanol, (2) leucine in a solvent mixtures of 1-butanol and dimethylformamide (DMF), and (3) leucine in a solvent mixtures of 1-butanol and N-methylpyrrolidone (NMP). The characteristic of the permeation cell accommodating the UDTR used to measuring real-time swelling and compaction is shown elsewhere [3].

In all the permeation experiments, 2.0 wt% leucine solutions at a feed rate of 40 ml/min were permeated in a transmembrane pressure range of 0.69–1.72 MPa. The membrane is allowed to swell in contact with the solution or solvents until equilibrium and membrane compacted simultaneous with permeation with the solutions. The swollen and compacted thicknesses are all measured in situ and real-time using UTDR. Note that the membranes were constrained during swelling. Here, most of the variables in Eq. 3.12 are known from experiments except K_d and the Poisson ratio (ν). If the membrane is sufficiently constrained, the Poisson ratio could be outside of the known ranges since the transverse strain as results of swelling and compaction could be far higher than the longitudinal strain. K_d can be calculated using Eqs. 2.1 and 2.2 (see Chap. 2, Sect. 2.2) where experimental data can be fitted to Eq. 2.1 and the unknown solute concentrations calculated from Eq. 2.2. UTDR is used to measure both swelling and compaction in real time. Here, only the reduced form of Eq. 3.12, i.e., $\psi = \frac{10^{-6}E\rho}{6(\delta_i-\delta_j)^2}\ln\left[\left(\frac{L_f}{L_c}\right)\left(\frac{1}{1+\alpha}\right)\right]$, is used for the prediction of C_{pi} since K_d and ν are unknown.

Table 4.12 presents the summary of the properties of the solvents, solvent mixture, and solute.

The membrane overall thickness is about 178 μm, and molar volume V_m is about 70,671 cm^3/mol.

$$\alpha = \frac{\overline{V}_0\lambda_f}{\overline{V}_m} \quad \overline{V}_i = \frac{M_w}{\rho} \quad \lambda_f = \% \text{ Swelling} \frac{\rho_{\text{solvent}}}{\rho_{\text{membrane polymer}}}$$

Table 4.12 Summary of the properties of the solvents and solute

Materials	Density (g/ml)	Solubility parameter (MPa$^{1/2}$)	$\Delta\delta$, i.e., ($\delta_i - \delta_j$)
1-butanol	0.81	23.1	3.60
60 wt% DMF and 40 wt% 1-butanol	0.89	24.1	4.60
50 wt% NMP and 50 wt% 1-butanol	0.92	23.0	3.50
Leucine (solute)	–	19.5	–
Starmem-122 (PI) membrane	–	26.6	–

$\Delta\delta|\delta_i - \delta_j|$, e.g., $23.1 - 19.5 = 3.60$; $24.1 - 19.5 = 4.60$; $23.0 - 19.5 = 3.50$

Table 4.13 Data from experiments [3]

Solvent	Average rejection (%)	Actual permeate solute C_{pi} (M)	V_o (ml/mol)	Swelling (%)	Swollen thickness (L_f) (μm)	Compacted thickness (L_c) (μm)
NMF/ butanol	46.2	0.029	89.9	70	125	155
DMF/ butanol	64.8	0.017	81.7	75	134	160
1-butanol	95.3	0.002	89.0	55	98	100

Swollen thickness (L_f) (μm), compacted thickness (L_c) (μm), permeate solute (C_{pi}) (M)

Table 4.14 Data from experiment and calculated information

Solvent	C_{pi} (M)	L_f (μm)	L_c (μm)	λ_f	E (MPa)	α	$\ln D$	$\Delta\delta^2$	ψ
NMF/ butanol	0.029	125	155	0.45	1.13	0.0006	−0.2189	11.6	0.0033
DMF/ butanol	0.017	134	160	0.47	1.13	0.0005	−0.1816	20.3	0.0015
1-butanol	0.002	98	100	0.31	1.60	0.0004	−0.0216	12.3	0.0004

Fig. 4.9 Experimental permeate solute concentration C_{pi} versus solvent densities

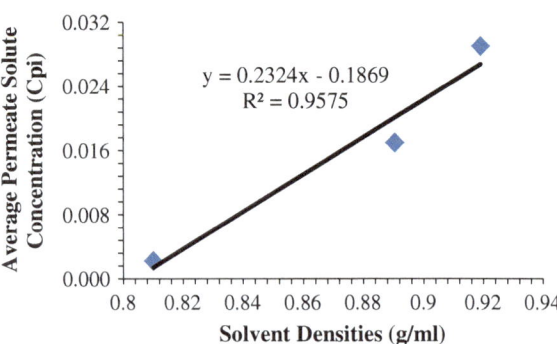

Table 4.13 shows the experimental data from permeating STARMEM-122 with the various solutions comprised of a solvent and mixture of solvents (see Chap. 3, Sect. 3.1).

Table 4.14 shows experimental and generated data for STARMEM-122 membrane permeation used for plotting the graph for the trend prediction

But $\ln D = \ln\left[\left(\frac{L_f}{L_c}\right)\left(\frac{1}{1+\alpha}\right)\right]$.

Figures 4.9 and 4.10 show the effects of the solvent densities on polyimide (PI) membrane performance. As expected, the PI membrane performances decreased as the solvent density increased.

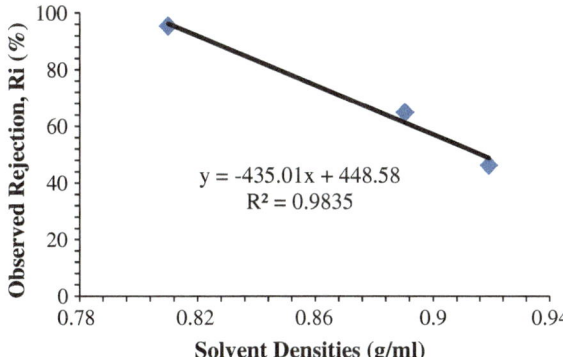

Fig. 4.10 Experimental solute observed rejection R_i (%) versus solvent densities

$$y = -435.01x + 448.58$$
$$R^2 = 0.9835$$

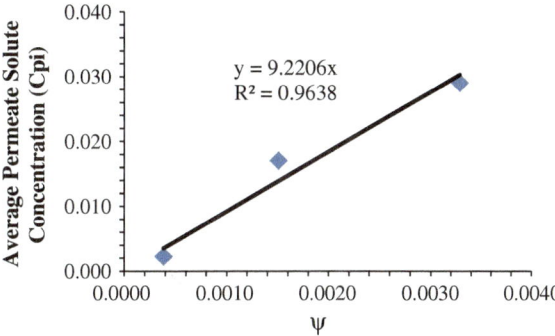

Fig. 4.11 Experimental permeate solute concentration C_{pi} versus predicted permeate solute concentration ψ showing expected direct relationship $\left(\psi = \frac{10^{-6}E\rho}{6(\delta_i - \delta_j)^2} \ln\left[\left(\frac{L_f}{L_c}\right)\left(\frac{1}{1+\alpha}\right) \right] \right)$

$$y = 9.2206x$$
$$R^2 = 0.9638$$

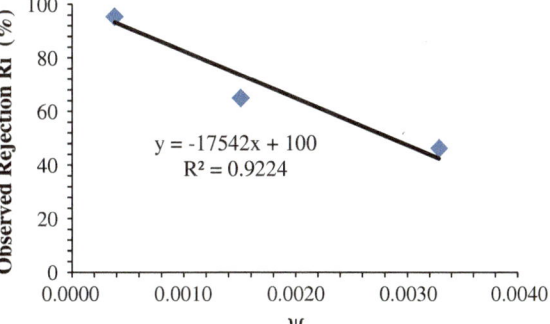

Fig. 4.12 Experimental solute observed rejection, R_i, (%) versus predicted permeate solute concentration ψ showing expected inverse relationship $\left(\psi = \frac{10^{-6}E\rho}{6(\delta_i - \delta_j)^2} \ln\left[\left(\frac{L_f}{L_c}\right)\left(\frac{1}{1+\alpha}\right) \right] \right)$

$$y = -17542x + 100$$
$$R^2 = 0.9224$$

In Fig. 4.9, the correlation between the experimental C_{pi} and the solvent density expected to be a straight line with a positive slope going through the origin (i.e., $y = mx$) had an intercept at 0.1869 (i.e., $y = mx + c$). This may be due to experimental error, or the systems may not be described by solvent densities alone.

Similarly, the plot in Fig. 4.10 calculated from Fig. 4.9 is expected to have a deviation.

Figure 4.10 i.e., the correlation between the experimental observed rejection and the solvent density used as prediction for Cpi expected to be a straight line with a negative slope and especially interception at 100 rather had an interception at 448.58. A similar explanation used for Fig. 4.9 can be applied here.

Figures 4.11 and 4.12 show plots of the actual PI performance with the predicted performance ψ. As expected, the membrane performances decrease with an increase in ψ values. Both Figs. 4.11 and 4.12 show the expected correlation between the predicted and experimental data and are all in accordance with the model prediction with no deviations.

References

1. Krea, M., Roizard, D., Moulai-Mustefa, N., & Sacco, D. (2004). Synthesis of polysiloxane-imide membranes—application to the extraction of organics from water mixtures. *Desalination, 163*, 203–206.
2. Czerwiński, W., Ostrowska-Gumkowska, B., Kozakiewicz, J., Kujawski, W., & Andrzej, W. (2004). Siloxane-urethane membranes for removal of volatile organic solvents by pervaporation. *Desalination, 163*, 207–214.
3. Anim-Mensah, A. R. (2007). Evaluation of solvent resistant nanofiltration (SRNF) membranes for small-molecule purification and recovery of polar aprotic solvents for re-use (Ph.D. thesis, University of Cincinnati, OH, 2007).

Chapter 5
Conclusions

Abstract The developed model was able to predict the trends for aqueous-organic and purely organic system separations. This section presents a summary of the key points in this book.

Keywords Performance prediction · Solution-diffusion · Organic solvents · Solvent-resistant · Swelling · Compaction · Polymer · Membranes · Membrane characterization · Solute distribution · Solubility parameter · Pervaporation · Nanofiltration · Separation · Purification · Membrane densification

- The developed model Eqs. 3.11 and/or 3.12 shows how some of the pertinent parameters for describing solution–diffusion polymeric membranes involved organic system separation and purification are correlated from a chemical, mechanical and thermodynamic standpoints and how they affect membrane performances.
- In the absence of data to use the developed model, reduced or simplified versions can be used for performance trend prediction.
- The model is able to predict membrane performance trends for both pervaporation and pressure-driven nanofiltration systems.
- Generally, as expected, excessive swelling or compaction results in lowering membrane separation performance. A balance between swelling and compaction is required to increase membrane performance.
- K_d which is the distribution or partitioning of the solute between the permeate and membrane (i.e., $K_d \leq C_{pi}/C_m$) did not have any marked significant effects on the membrane separation performance within the limits used to test the model.
- Generally, membrane densification as results of compaction while membrane is permeated and well-constrained improved membrane performance.
- $\alpha = \frac{V_0 \lambda_f}{V_m}$ which is the mass ratio of solvent retained in a membrane to the dry weight of the membrane, did not have significant effects on the separation

© The Author(s) 2015

A. Anim-Mensah and R. Govind, *Prediction of Polymeric Membrane Separation and Purification Performances*, SpringerBriefs in Molecular Science, DOI 10.1007/978-3-319-12409-4_5

performance, this is because $\alpha <<< 1$, because $\overline{V_m} >>>> \overline{V_o}$ or $\frac{1}{1+\alpha} \to 1$. $\overline{V_m}$—membrane polar molar volume and $\overline{V_o}$—membrane polar molar volume.

- A balance is expected between swelling and compaction to achieve better separation performance. Extreme swelling and extreme compaction each is negative on performances.
- The effect of combined swelling and compaction is determined from the ratio of swelling to compaction (L_f/L_c). A decrease in the ratio results in an increase in membrane performance. A decrease in the ratio may mean low L_f value at high L_c during permeation. This means excessive compaction as already known could lead to poor membrane performance; hence, sufficient compaction at reasonable swelling is preferred.
- Low compressive Young's modulus (E) dominates membrane swelling; however, low E could be a result of the interaction between the membrane material and the solvent in which the interaction determines the extent of swelling. Hence, some reasonable amount of swelling is required to improve membrane performance.
- In the absence of experimental information, the solvent or solvent mixtures densities (ρ_i), the ratio of the solvent density to the square of the solubility parameter difference between the solvent and solute, i.e., $\rho_i/\Delta\delta^2$, and $\psi = \frac{10^{-6}E\rho}{6(\delta_i-\delta_j)^2} \ln\left[\left(\frac{L_f}{L_c}\right)\left(\frac{1}{1+\alpha}\right)\right]$ could all be used to predict membrane separation performance trend. However, the results obtained in this book suggest the reliability in the order of ψ, $\rho_i/\Delta\delta^2$, and then ρ_i. The density (ρ_i) prediction could breakdown when used for organic solvents of the same group, i.e., either non-polar, polar protic, or polar aprotic because in the same group other factors may dominate other than the effects of the polar, dispersion, or hydrogen bonding contributions on density.
- The solvent density appears to predict the trends for both the pervaporation and nanofiltration membrane separations more importantly for the below trend.

Polar Aprotic > Polar Protic > Non-Polar

(see Table 2.1, Sect. 2.1 and Chap. 2). However, the ratio of the density to the square of the solubility parameter difference between the solvent and solute (i.e., $\rho_i/\Delta\delta^2$) was sufficient to predict the trends for the pervaporation membrane separations while $\psi\left(\frac{10^{-6}E\rho}{6(\delta_i-\delta_j)^2} \ln\left[\left(\frac{L_f}{L_c}\right)\left(\frac{1}{1+\alpha}\right)\right]\right)$ for the nanofiltration membrane. This means more parameters will be required to describe the nanofiltration membrane than the pervaporation membrane separations. This could be due to the nature of the feed solution involved i.e., solvent mixtures for pervaporation separation versus a solute in a solvent for nanofiltration separation, the maximum transmembrane pressure and temperature involved.

Chapter 6
Future Directions

Abstract This chapter presents the future directions of developed model to new areas and extension of the model to increase the versatility to elucidate some of the underlying principles and provide in-depth understanding. In addition more consideration will be given to the newly defined dimensionless number interpretation and its application.

Keywords Dimensionless parameter · Organic systems · Aqueous-organic · Purely organic · Flux · Feed flowrate · Pore-flow · Microfiltration · Ultrafiltration

The model developed here is only preliminary and only considers membrane rejection; however, there are several opportunities to improve on the model to provide understanding to some of the complex issues related to the application of solution–diffusion membranes for organic solvents' environments. The new dimensionless parameter defined in this book which is characteristic of the separation system Eq. 3.14 (see Chap. 3, Sect. 3.1) will be given the attention for universal application solution–diffusion membranes. The model will be tested on other published organic separation systems to provide explanations on some of the contributions to the observed trends. Also, K_d variations for different published organic separation systems will be investigated if available for the different organic separation system. Moreover, extending the model to consider flux prediction and effects of feed flow rate, bulk solute feed concentrations and transmembrane pressure for predicting both membrane rejection and flux will be necessary. Finally, the model will be extended to consider pore-flow membranes such as microfiltration and ultrafiltration.

© The Author(s) 2015
A. Anim-Mensah and R. Govind, *Prediction of Polymeric Membrane Separation and Purification Performances*, SpringerBriefs in Molecular Science,
DOI 10.1007/978-3-319-12409-4_6